Test Yourself

Organic Chemistry

Drew H. Wolfe, Ph.D.
Professor of Chemistry
Hillsborough Community College
Tampa, FL

Contributing Editors

Peter K. Trumper, Ph.D.
Department of Chemistry
University of Southern Maine
Portland, ME

Edward J. Walsh, Ph.D.
Department of Chemistry
Allegheny College
Meadville, PA

Christopher J. Cramer, Ph.D.
Department of Chemistry and Chemical Physics
Institute of Technology
University of Minnesota
Minneapolis-St. Paul, MN

NTC LEARNINGWORKS
a division of NTC Publishing Group
Lincolnwood, Illinois

Library of Congress Cataloging-in-Publication Data
is available from the Library of Congress

A *Test Yourself Books, Inc.* Project

Published by NTC Publishing Group
© 1996 NTC Publishing Group, 4255 West Touhy Avenue
Lincolnwood (Chicago), Illinois 60646-1975 U.S.A.

Contents

Preface

Test Yourself in Organic Chemistry has been written to help organic chemistry students achieve higher grades on quizzes, exams, midterms, and final exams. *Organic Chemistry* covers most of the principal topics in a first-year organic chemistry course. The main topics covered in this book include: hydrocarbons, hydrocarbon derivatives, stereochemistry, reaction mechanisms, synthetic methods, and spectroscopy.

Each chapter in this book has four parts. The first part, Brief Yourself, is a brief discussion on the topic, including important definitions, reactions, structures, and mechanisms. The second part, Test Yourself, consists of both multiple choice and open-format questions and problems that typically appear on organic chemistry exams. You can either answer all of the questions or the ones emphasized in your course. To help you decide which questions you should try, refer to the fourth section, Grade Yourself, which lists the major topics of the chapter along with the questions that fall into each category. After completing the practice test, check your answers against those in the third section; Check Yourself. Besides the correct answers, this section also shows equations, structures, synthetic pathways, and mechanisms. When you finish, fill out the Grade Yourself key, circling the numbers of the questions that you answered incorrectly. After reviewing these items, you should go back to your textbook and notes to try to better understand these topics. You might want to try these incorrectly answered questions again so that you will not make the same mistakes when you take the actual test.

In conclusion, I would like to thank Fred Grayson, who has given me the opportunity to write this and other books.

Drew H. Wolfe, Ph.D.

How to Use this Book

This "Test Yourself" book is part of a unique series designed to help you improve your test scores on almost any type of examination you will face. Too often, you will study for a test—quiz, midterm, or final—and come away with a score that is lower than anticipated. Why? Because there is no way for you to really know how much you understand a topic until you've taken a test. The *purpose* of the test, after all, is to test your complete understanding of the material.

The "Test Yourself" series offers you a way to improve your scores and to actually test your knowledge at the time you use this book. Consider each chapter a diagnostic pretest in a specific topic. Answer the questions, check your answers, and then give yourself a grade. Then, and only then, will you know where your strengths and, more importantly, weaknesses are. Once these areas are identified, you can strategically focus your study on those topics that need additional work.

Each book in this series presents a specific subject in an organized manner, and although each "Test Yourself" chapter may not correspond exactly to the same chapter in your textbook, you should have little difficulty in locating the specific topic you are studying. Written by educators in the field, each book is designed to correspond, as much as possible, to the leading textbooks. This means that you can feel confident in using this book, and that regardless of your textbook, professor, or school, you will be much better prepared for anything you will encounter on your test.

Each chapter has four parts:

 **Brief Yourself.** All chapters contain a brief overview of the topic that is intended to give you a more thorough understanding of the material with which you need to be familiar. Sometimes this information is presented at the beginning of the chapter, and sometimes it flows throughout the chapter, to review your understanding of various *units* within the chapter.

 **Test Yourself.** Each chapter covers a specific topic corresponding to one that you will find in your textbook. Answer the questions, either on a separate page or directly in the book, if there is room.

 **Check Yourself.** Check your answers. Every question is fully answered and explained. These answers will be the key to your increased understanding. If you answered the question incorrectly, read the explanations to *learn* and *understand* the material. You will note that at the end of every answer you will be referred to a specific subtopic within that chapter, so you can focus your studying and prepare more efficiently.

 **Grade Yourself.** At the end of each chapter is a self-diagnostic key. By indicating on this form the numbers of those questions you answered incorrectly, you will have a clear picture of your weak areas.

There are no secrets to test success. Only good preparation can guarantee higher grades. By utilizing this "Test Yourself" book, you will have a better chance of improving your scores and understanding the subject more fully.

Introduction to Organic Chemistry

Brief Yourself

Organic chemistry is the study of the properties, structures, and reactions of most carbon compounds. It is one of the most exciting areas of chemistry because of the importance of organic compounds. Carbon compounds are found in living things and compose most of the things that we own.

The two major groups of organic compounds include the hydrocarbons and hydrocarbon derivatives. Hydrocarbons are compounds that contain C and H. Hydrocarbon derivatives contain C, usually H, and at least one other element, e.g., O, N, S, P, or a halogen, X (F, Cl, Br, or I).

The four main classes of hydrocarbons are the alkanes, alkenes, alkynes, and aromatics. The alkanes are hydrocarbons with C–C single bonds. Alkenes are hydrocarbons that have a C–C double bond, and alkynes have a C–C triple bond. Aromatic hydrocarbons have a special stable structure that consists of a planar ring with alternating double and single bonds. A good example of an aromatic hydrocarbon is benzene, C_6H_6. Hydrocarbon molecules can be unbranched chains of C atoms, branched chains of C atoms, rings of C atoms, or a combination of chains, branched chains, or rings.

Many different groups of hydrocarbon derivatives exist because C atoms can bond to atoms such as O, S, N, P, and X in many different ways. All hydrocarbon derivatives have an organic component called an R group, which can be composed of any number of C atoms, and one or more functional groups. The functional group is an atom or group of atoms that gives the compound its characteristic physical and chemical properties. Some common hydrocarbon derivatives that contain O are alcohols, R–OH; ethers, ROR′; aldehydes, RCHO; ketones, RCOR′; carboxylic acids, RCOOH; and esters, RCOOR′. A common hydrocarbon derivative that contains N is the primary amines, RNH_2. A group that contains both O and N atoms is the amides, RCONHR′. The group of hydrocarbon derivatives that have halogen atoms is called the halogenated hydrocarbons, RX.

To understand organic molecules, a good basic understanding of the principles of chemical bonding is necessary. The two fundamental types of bonds are ionic and covalent bonds. Ionic bonds result when one or more electrons is transferred from an atom or group of atoms to another atom or group of atoms. Examples of ionic compounds include sodium chloride, Na^+Cl^-; ammonium nitrate, $NH_4^+NO_3^-$; and potassium acetate, $K^+CH_3COO^-$. Covalent molecules result when electron clouds (orbitals) from one nonmetal atom overlap with an electron cloud from another nonmetal atom. Some people say that the electrons are "shared" in covalent bonds because one nucleus attracts one or more electrons from the other nonmetal atom. Most organic compounds are essentially covalent molecules.

If two electrons are "shared" between two nonmetal atoms, a single covalent bond results. If four or six electrons are "shared," then a double or triple bond, respectively, results. The number of bonds between two nonmetal atoms is called the bond order. For example, a bond order of 2 means a double covalent bond and a bond order of 3 means a triple covalent bond.

Other bond characteristics include the bond length, bond energy, and bond polarity. The bond length is the average distance between two nuclei that form a covalent bond. This distance is usually expressed in picometers, pm, or nanometers, nm. As the bond order increases, the bond length decreases. The bond energy is the amount of energy needed to break the bond homolytically, which means that each atom obtains the same number of electrons. Higher bond energies mean stronger bonds. As the bond order increases, the bond energy increases. If the electrons are shared equally, a nonpolar bond results. If the electrons are shared unequally, a polar bond results. The magnitude of the polarity of the bond depends on the difference in electronegativity, ΔEN. The dipole moment, μ, is the experimental measurement of the polarity of a molecule. If the dipole moment is zero, then the molecule is nonpolar, and if the dipole moment is greater than zero, then the molecule is polar.

Lewis structures are used to show how organic molecules are bonded. The most stable Lewis structures are those in which all atoms obtain the stable noble gas configuration. From the Lewis structure and applying the valence shell electron pair repulsion method (VSEPR method), the structures of simple organic compounds can be predicted. The principal geometries associated with organic molecules are linear (two bonds and no lone pairs), trigonal planar (three bonds and no lone pairs), trigonal pyramidal (three bonds and one lone pair), angular (two bonds and two lone pairs), and tetrahedral (four bonds and no lone pairs).

Some organic molecules have more than one correct Lewis structure. These molecules exhibit resonance. Resonance is found in molecules with electron delocalization. This means that some of the electrons in bonds are spread out over more than two atoms. Molecules that exhibit resonance tend to be more stable than similar molecules that do not.

An important bonding concept in organic chemistry is formal charge. The formal charge of an atom in a molecule shows the electron environment of that atom in the molecule. If it has a positive formal charge, it has lost some of its electron density. If it has a negative formal charge, it has gained electron density. If it has a zero formal charge, it has essentially the same electron density as it does in the unbonded atom. The formal charge of an atom is calculated as follows.

Formal charge = number of valence electrons – [(number of electrons in bonds/2) +

(number of electrons in lone pairs)]

For example, the formal charge on the O atom in H_3O^+ is +1 (Formal charge = $6 - [(6/2) + 2] = +1$).

Organic chemists have different ways of showing the Lewis structures of organic molecules. For example, the Lewis structure of ethane, C_2H_6, can be written as follows:

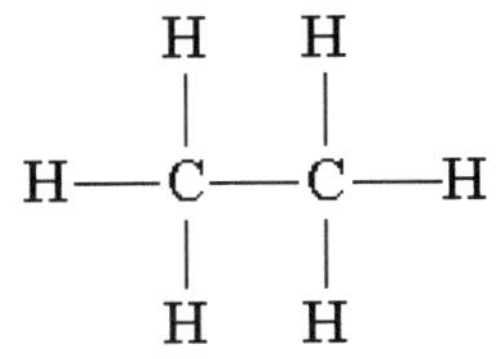

H   H
| |
H—C—C—H
| |
H   H

ethane

In this Lewis structure each line is a shared pair of electrons. A more convenient way to express the structure of ethane is as follows.

$$CH_3–CH_3$$
ethane

Using this method, each C atom is written followed by the number of H atoms that bond to it. Only the C–C bonds are shown in this structure. It can be shortened to just show the atoms without any bonds.

$$CH_3CH_3$$
ethane

Another way to represent organic structures is to eliminate all of the atoms and only show the bonds. At each intersection and terminus is a C atom. For example, octane, C_8H_{18}, can be represented as follows.

octane

If the C and H atoms are included in the structure, octane appears as follows.

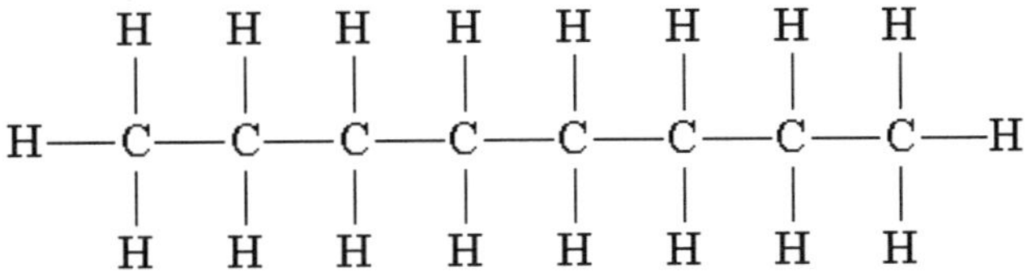

H   H   H   H   H   H   H   H
| | | | | | | |
H—C—C—C—C—C—C—C—C—H
| | | | | | | |
H   H   H   H   H   H   H   H

octane

Removing the lines that represent the C–H bonds from octane produces the following.

$$CH_3-CH_2-CH_2-CH_2-CH_2-CH_2-CH_2-CH_3$$

octane

Finally, this structure can be further condensed by placing the repeating methylene units, CH_2, in parentheses and adding a subscript that tells how many are in the chain.

$$CH_3-(CH_2)_6-CH_3$$

octane

Constitutional isomers of organic compounds are often found. Constitutional isomers are compounds that have the same molecular formula but have a different structural formula. This means that they have the same number of each type of atom, but the way they are bonded is different. In other words, these compounds have different constitutions. Two constitutional isomers are ethanol and dimethyl ether, C_2H_6O.

$$CH_3-CH_2-OH \qquad\qquad CH_3-O-CH_3$$

ethanol   dimethyl ether

Test Yourself

1. a. What are the two principal groups of organic compounds?
    b. What distinguishes these two groups?

2. a. What is the general formula for all alcohols?
    b. What is the Lewis structure for the simplest alcohol?

3. What are the names of the four classes of hydrocarbons?

4. What is the general formula for all aldehydes?

5. Write the formulas for the simplest halogenated hydrocarbons.

6. What is the formal charge on C in the following chemical species?

$$H-\overset{\cdot\cdot}{\underset{|}{C}}-H$$
$$H$$

7. Arrange the following bonds from lowest to highest polarity: C–F, C–C, and C–Cl.

8. Draw the Lewis structure for hydrogen cyanide, HCN, and determine the bond order of the C–N bond.

9. a. Which of the following should have the highest bond energy: C–C single bond, C–C double bond, or C–C triple bond?
    b. Which of these bonds is the shortest? Explain.

10. Consider the following anion.

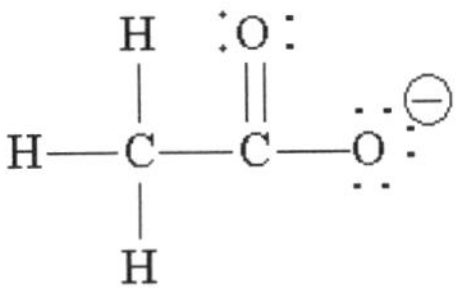

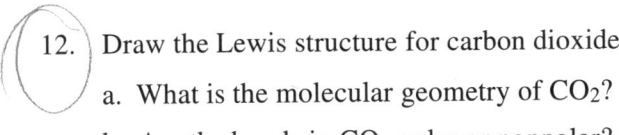

Draw a resonance structure of this anion.

11. What is the molecular geometry of boron trifluoride, BF_3?

12. Draw the Lewis structure for carbon dioxide.

    a. What is the molecular geometry of CO_2?

    b. Are the bonds in CO_2 polar or nonpolar?

    c. Is the CO_2 molecule polar or nonpolar?

13. Simplify the following structure to one that only uses lines.

$$CH_3-(CH_2)_2-\underset{\underset{Cl}{|}}{CH}-CH_3$$

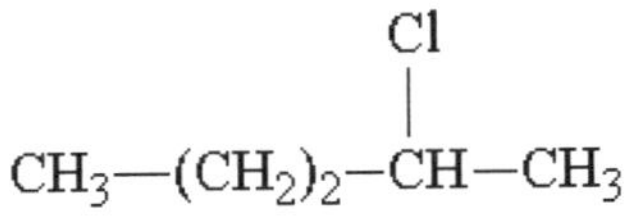

14. Simplify the following structure to one that only uses lines.

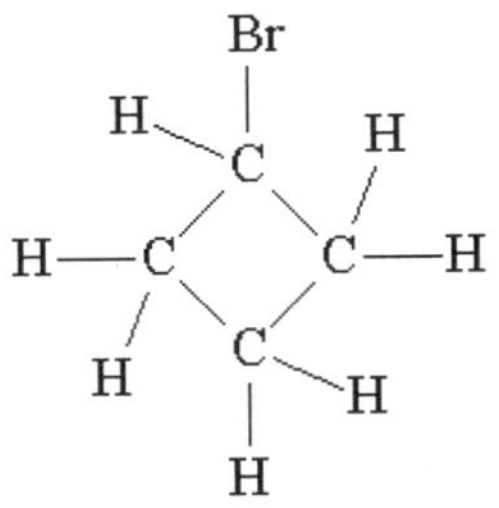

15. Draw the complete Lewis structure for the following.

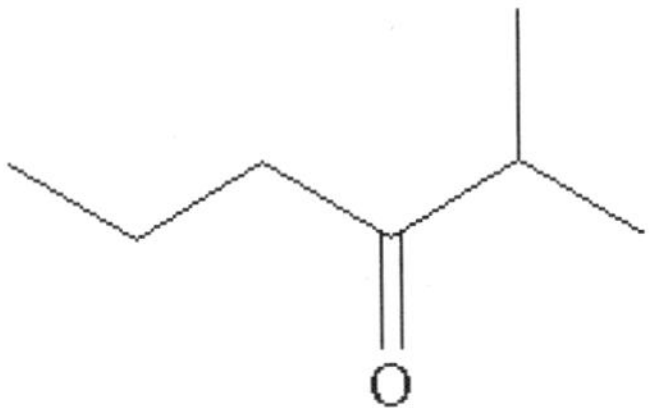

16. Draw the complete Lewis structure for the following.

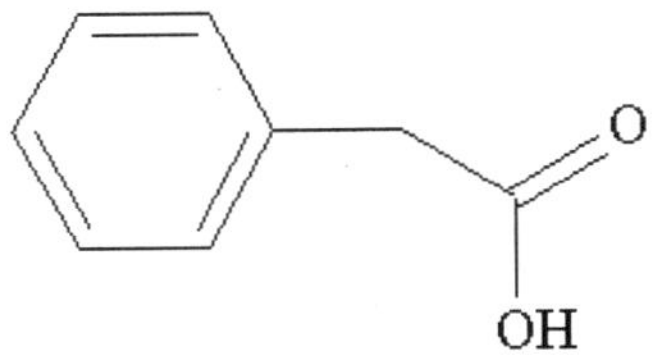

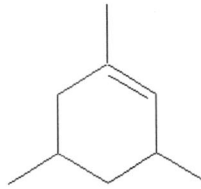

17. Draw a constitutional isomer of 1-propanol, $CH_3CH_2CH_2OH$.

18. Draw the structural formulas, using lines only, for three compounds that have the molecular formula of C_4H_8 and do not have a ring structure.

19. Draw two constitutional isomers of $C_2H_4Cl_2$.

20. What is the molecular formula for the following compound?

21. Which of the following has a zero dipole moment?

    a. CH_4

    b. CI_4

    c. CO_2

    d. all of these

22. Which of the following is an amide?

    a. ROR′

    b. RCOR′

    c. RNH_2

    d. $RCONH_2$

(Consider the following structures to answer questions 23 to 25)

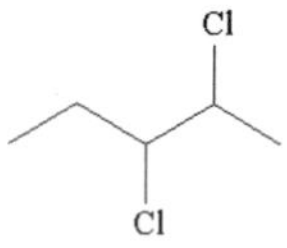

CH$_3$—CH$_2$—CH—CH—CH$_3$ with Cl Cl on the two middle carbons

I 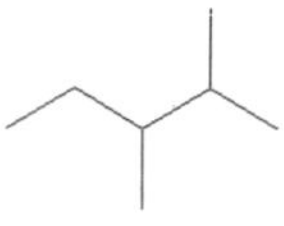   II 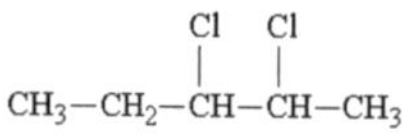   III

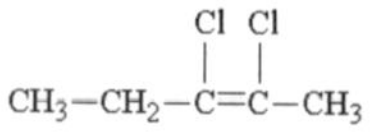

CH$_3$—CH$_2$—C=C—CH$_3$ with Cl Cl

IV

H—C—C—C—C—C—H with H H CH$_3$ CH$_3$ H on top and H H H H H on bottom

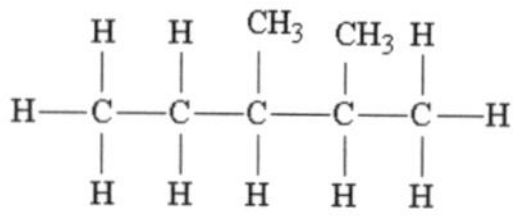

V

23. Which of these structures represent the same molecules?

   a. I and II

   b. I and IV

   c. II and V

   d. I and III, II and V

24. Which of these structures is classified as an alkene?

   a. I

   b. II

   c. III

   d. IV

25. Which of these structures has the lowest molecular mass (weight)?

   a. I and IV

   b. II and V

   c. III

   d. IV

   e. none of these

26. What is the formal charge on the C atom in the following structure?

   Cl——C——Cl
        |
        Cl

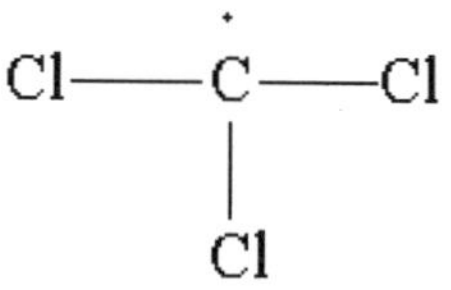

   a. 0

   b. −1

   c. +1

   d. none of these

27. What is the general formula for all carboxylic acids?

   a. RCOR

   b. ROR

   c. RCOOH

   d. RCOOR

28. Draw the Lewis structure of formaldehyde, H$_2$CO. What is the bond order of the C–O bond?

   a. 0

   b. 1

   c. 2

   d. 3

29. What is the formal charge on the O atom in the following structure?
   $CH_3CH_2OH_2$

   a. 0

   b. +1

   c. −1

   d. +2

30. Which is a constitutional isomer of the following?

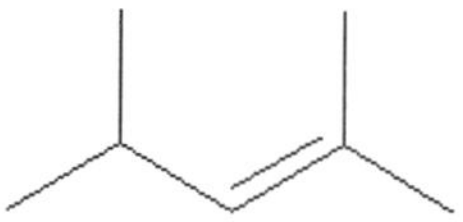

   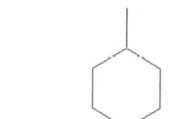

   a.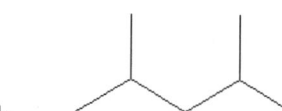

   b.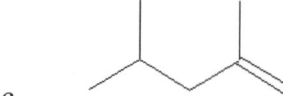

   c.

   d. a and c

Check Yourself

1. a. Hydrocarbons and hydrocarbon derivatives are the two principal groups of organic compounds.
    b. Hydrocarbons only have C and H atoms. Hydrocarbon derivatives have C, usually H, and some other atom such as O, N, S, P, or Cl. **(Hydrocarbons and hydrocarbon derivatives)**

2. a. The general formula for all alcohols is ROH.
    b. The simplest alcohol has one C atom; thus, its molecular formula is CH_3OH. The Lewis structure for CH_3OH, methanol, is

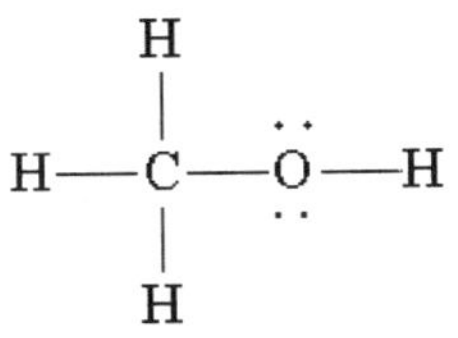

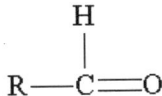

methanol

(Hydrocarbon derivatives)

3. Alkanes, alkenes, alkynes, and aromatics are the four groups of hydrocarbons. **(Hydrocarbons)**

4. a. Aldehydes have the general formula of RCHO.
    b. The Lewis structure for aldehydes is:

$$R-\overset{\overset{\displaystyle H}{|}}{C}=O$$ **(Hydrocarbon derivatives)**

5. CH_3F, CH_3Cl, CH_3Br, and CH_3I **(Halogenated hydrocarbons)**

6. Formal charge (C) $= 4 - [(6/2) + 2] = -1$ **(Formal charge)**

7. C–C is nonpolar and thus has the lowest polarity. Because F has a higher electronegativity than Cl, the C–F bond is more polar than C–Cl. Therefore, the ranking is as follows.

    C–C<C–Cl<C–F

    (Chemical bonds)

8. The Lewis structure for HCN is

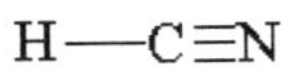

    $$H-C\equiv N$$

    Hence, the C–N bond order is 3. **(Chemical bonds)**

9. a. A bond with a higher bond order has a higher bond energy; thus, the C–C triple bond would have the highest bond energy.
    b. A bond with a higher bond order is shorter than one with a lower one; thus, the C–C triple bond is the shortest. **(Chemical bonds)**

10. Move one of the lone pairs on the O atom with the negative charge to form a C–O double bond. This would change one bond of the double bond to a lone pair.

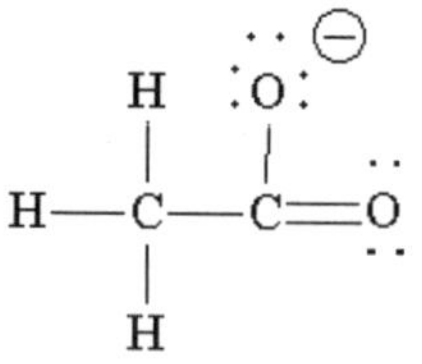

    (Resonance)

11. Drawing the Lewis structure for boron trifluoride shows that the central B atom has three B–F single bonds with no lone pairs; thus, the molecular geometry for BF_3 is trigonal planar. **(Chemical bonds)**

12. The Lewis structure for CO_2 is as follows.

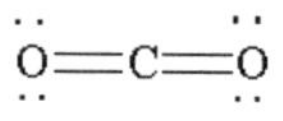

    a. Because the C atom has two double bonds to the O atoms and no lone pairs, the molecule is linear.
    b. Each C–O bond is polar because the O atom is more electronegative than the C atom.
    c. The overall molecule is nonpolar because the two dipoles are equal but opposite and they cancel.
    (Chemical bonds)

13.

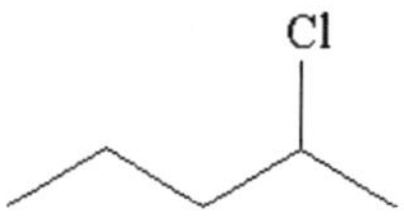

    (Organic structures)

14.

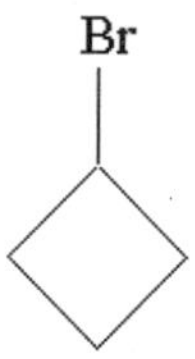

    (Organic structures)

15.

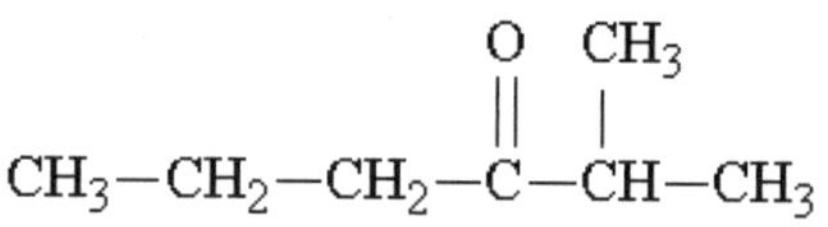

    (Organic structures)

16.

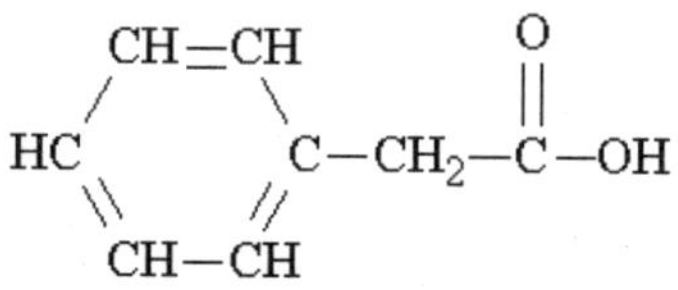

    (Organic structures)

17. CH_3CH_2–O–CH_3 or $CH_3CH(OH)CH_3$ **(Isomers)**

18.

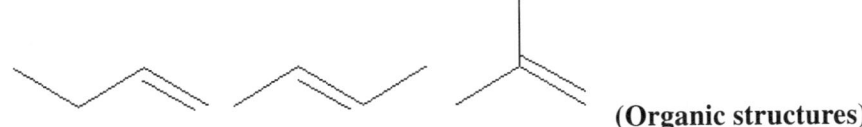 **(Organic structures)**

19.

$$CH_3-\overset{\overset{\displaystyle Cl}{|}}{CH}-Cl \qquad Cl\!-\!CH_2-CH_2-Cl$$

(Isomers)

20. The molecular formula for this compound is C_9H_{16}. **(Molecular formulas)**

21. d. All of these compounds are nonpolar; hence, their dipole moments equal zero. **(Chemical bonding)**

22. d. $RCONH_2$ is the general formula for a primary (1°) amide. **(Hydrocarbon derivatives)**

23. d. I and III both have chains of five C atoms with Cl atoms on the second and third C atoms. II and V both have two CH_3 groups (methyl groups) bonded to the second and third C atoms of a chain of five C atoms. **(Organic structures)**

24. d. IV is the only compound that has a double bond. All alkenes have a double bond. **(Hydrocarbons)**

25. b. II and V are different ways of showing the same structure. This molecule has the lowest molecular mass. Its formula is C_7H_{16}, which means it molecular mass is 100. Structure I is the same except that two CH_3 with a total mass of 30 are replaced with two Cl atoms with a mass of 71. **(Molecular formulas)**

26. a. Formal charge (C) = $4 - [(6/2) + 1] = 0$ **(Formal charge)**

27. c. RCOOH is the general formula for all carboxylic acids (organic acids). **(Hydrocarbon derivatives)**

28. c. The Lewis structure of formaldehyde shows a C–O double bond and two C–H single bonds. **(Chemical bonds)**

29. b. Formal charge (O) = $6 - [(6/2) + 2] = +1$ **(Formal charge)**

30. d. Both a and c have the formula of C_7H_{14}; thus, they are constitutional isomers of the original structure. **(Isomers)**

Grade Yourself

Circle the number of questions you missed, then fill in the total incorrect for each topic. If you answered more than three questions incorrectly, you need to focus on that topic. (If a topic has less than three questions and you had at least one wrong, we suggest you study that topic also. Read your textbook, a review book, or ask your teacher for help.)

Subject: Introduction to Organic Chemistry

Topic	Question Numbers	Number Incorrect
Hydrocarbon and hydrocarbon derivatives	1	
Hydrocarbons	3, 24	
Hydrocarbon derivatives	2, 4, 22, 27	
Halogenated hydrocarbons	5	
Formal charge	6, 26, 29	
Chemical bonds	7, 8, 9, 11, 12, 21, 28	
Resonance	10	
Organic structures	13, 14, 15, 16, 18, 23	
Isomers	17, 19, 30	
Molecular formulas	20, 25	

Alkanes and Cycloalkanes

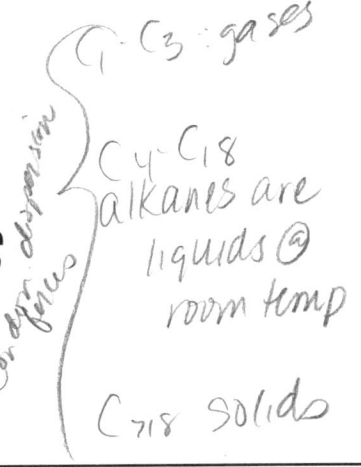

2

Brief Yourself

Alkanes are a homologous series of compounds with the general formula of C_nH_{2n+2}. A homologous series is a group of compounds with the same general formula in which the next member of the series only differs from the previous one by a "CH_2" unit. Members of a homologous series have similar chemical properties and exhibit a predictable range of physical properties.

The simplest alkane is the gas methane, CH_4. The next two members of the series, also gases, are ethane, C_2H_6, and propane, C_3H_8. The alkanes with four to ten C atoms are butane, C_4H_{10}; pentane, C_5H_{12}; hexane, C_6H_{14}; heptane, C_7H_{16}; octane, C_8H_{18}; nonane, C_9H_{20}; and decane, $C_{10}H_{22}$. These compounds are liquids at room conditions.

The structures of the alkanes resemble that of the simplest compound, methane. The Lewis structure for methane is

$$
\begin{array}{c}
H \\
| \\
H - C - H \\
| \\
H
\end{array}
$$

methane

The C atom in methane has four sp^3 hybrid orbitals that overlap with the $1s$ orbitals of H atoms. Associated with the sp^3 hybrid orbital is tetrahedral geometry and 109.5° bond angles. Thus, methane and the other alkanes are nonpolar molecules because the four slightly polar C–H bonds cancel as a result of the symmetrical tetrahedral structure.

The principal intermolecular forces between alkane molecules are London dispersion forces. Due to these weak forces the simplest alkanes, C_1 to C_3, are gases, C_4 to C_{18} are liquids, and those with more than 18 carbons are solids at 298 K.

The International Union of Pure and Applied Chemistry, IUPAC, has established the rules to name the alkanes and all other organic compounds. Many simple alkanes also have common (sometimes called trivial) names. For example, the unbranched molecule with the formula C_5H_{12} is called pentane in the IUPAC system and *n*-pentane in the common name system. In the common naming system, the *n* stands for normal or unbranched. This distinguishes it from one of the isomers of n-pentane. For example, let's consider the following molecule and how it is named in the two systems.

$$\text{or} \qquad CH_3-CH_2-\underset{\underset{CH_3}{|}}{CH}-CH_3$$

Note that two different ways are used to draw the same structure. One uses only lines, and the other places the C atoms in the chain along with the number of H atoms bonded to them. In the IUPAC system the longest continuous chain is first found and then the substituent groups are identified and located on the chain. In this molecule the longest chain is four C atoms; thus, the stem name is "butane." Located on the second C atom from the end of the chain is a methyl group, CH_3. Therefore, the IUPAC name for this molecule is 2-methylbutane. The common name for this compound is isopentane. The "iso" means that a methyl group is branched at the second carbon atom. Other prefixes in the common naming system are *sec*, *tert*, and *neo*. Consider the following compounds listed with both their IUPAC and common names.

$$CH_3-\underset{\underset{CH_3}{|}}{CH}-CH_3 \qquad\qquad CH_3-\underset{\underset{CH_3}{\overset{\overset{CH_3}{|}}{|}}}{C}-CH_3$$

2-methylpropane   2,2-dimethylpropane

(isobutane)   (neopentane)

Alkyl groups are substituent groups bonded to C chains or functional groups. An alkyl group results when a hydrogen atom is removed from an alkane. To write the name of an alkyl group, the *ane* ending for the alkane is removed and is replaced by *yl*. Examples of simple alkyl groups are CH_3-, methyl; CH_3CH_2-, ethyl; and $CH_3CH_2CH_2-$, propyl. More complex alkyl groups are listed below.

$$CH_3-\underset{\overset{|}{CH_3}}{CH}- \qquad\qquad CH_3-\underset{\underset{CH_3}{|}}{\overset{\overset{CH_3}{|}}{C}}- \qquad\qquad CH_3-CH_2-\underset{\overset{|}{CH_3}}{CH}-$$

1-methylethyl   1,1-dimethylethyl   1-methylpropyl

(isopropyl)   (t-butyl)   (sec-butyl)

If two or more alkyl groups are bonded to a chain, they are identified and the numbers of the C atoms to which they are attached are included in the name. If more than one of the same alkyl group are attached to the chain the prefixes di-, tri-, tetra-, and penta- are used. Consider the following example of writing the IUPAC name of a more complex alkane.

$$CH_3$$
$$|$$
$$CH_3 \quad CH_2 \quad CH_3$$
$$| \quad \quad | \quad \quad |$$
$$CH_3{-}(CH_2)_3{-}CH{-}CH{-}C{-}CH_3$$
$$|$$
$$CH_3$$

3-ethyl-2,2,4-trimethyloctane

This molecule has the general formula of $C_{13}H_{28}$; thus, it is a constitutional isomer of tridecane.

Alkanes tend to be unreactive. One important reaction that the alkanes undergo is combustion—they burn in the presence of oxygen, O_2. For example, the combustion of methane with excess oxygen present can be represented as follows.

$$CH_4 + 2O_2 \rightarrow CO_2 + 2H_2O$$

The enthalpy of combustion, ΔH_{comb}, is the amount of energy released when an organic compound is combusted.

Structures that only differ by rotation around a single bond are different conformations. The study of these different forms that a molecule can take is called conformational analysis.

Newman projections are used to show different conformations of a molecule. For example, the Newman projections for the most important staggered conformers of butane sighting along the C_2–C_3 bond are as follows.

Anti conformation   Gauche conformation
butane   butane

The torsional angle (also called the dihedral angle) between the methyl groups in the anti conformation of butane is $180°$, which is larger than the $60°$ torsional angle in the gauche conformation of butane. The anti conformation has less torsional strain than the gauche conformation because the methyl groups are more widely separated, which minimizes the repulsive forces in the molecule. The gauche conformation of butane is 0.9 kcal (3.8 kJ) higher in energy than the anti conformation because of the increased repulsive forces between the methyl groups, which are closer together in the gauche conformation.

The energies for the eclipsed conformations of butane are higher than those of the staggered conformations because the three sets of groups are as close to each other as possible. The two principal eclipsed conformations of butane are listed below with their torsional energies compared to the anti staggered conformation of butane.

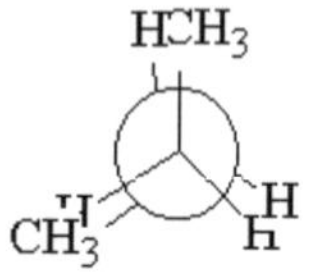

Eclipsed conformation
butane
(3.6 kcal, 15 kJ)

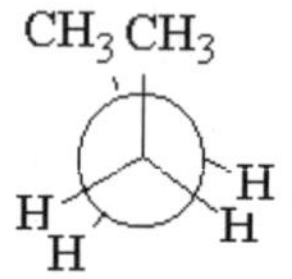

Totally eclipsed conformation
butane
(6 kcal, 25 kJ)

Whenever two bulky groups repel and increase the energy of the system, it is termed steric strain.

Cycloalkanes differ from the other alkanes in that they have a ring structure. As a result, their general formula is C_nH_{2n}. The simplest cycloalkanes are cyclopropane, C_3H_6; cyclobutane, C_4H_8; cyclopentane, C_5H_{10}; and cyclohexane, C_6H_{12}.

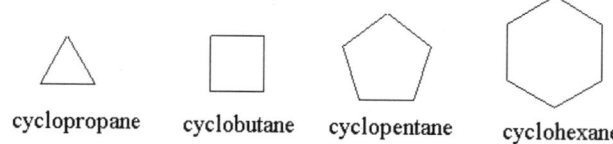

cyclopropane   cyclobutane   cyclopentane   cyclohexane

To write the name of a substituted cycloalkane, the stem name is the name of the cycloalkane ring. The substituent groups are numbered and named in a manner similar to the other alkanes. Consider the following examples of substituted cycloalkanes.

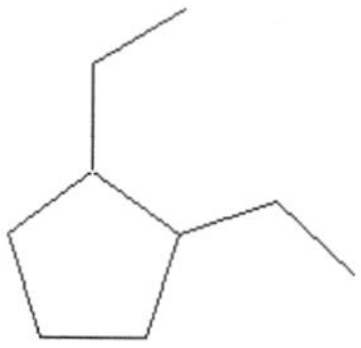

1,2-diethylcyclopentane

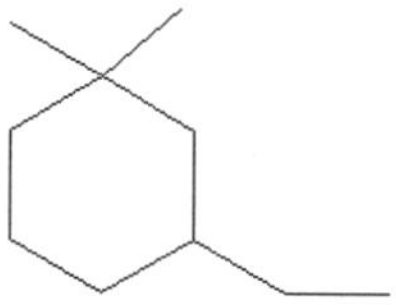

3-ethyl-1,1-dimethylcyclohexane

Sometimes a ring is a substituent group bonded to a chain or another ring. For this case, the ring is an alkyl group. Examples include cyclopropyl, cyclobutyl, and cyclopentyl. Two examples of such molecules follow.

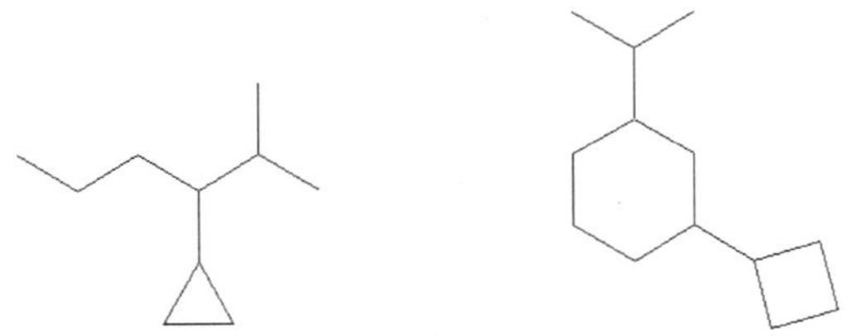

3-cyclopropyl-2-methylhexane   1-isopropyl-3-cyclobutylcyclohexane

The cycloalkane ring is rigid and does not allow freedom of rotation as in the other alkanes. Hence, two groups attached to a cycloalkane ring can either be on the same or opposite sides of the ring. If the two groups are on the same side, they are *cis* to each other, and if they are on opposite sides, they are *trans* to each other. Molecules that have the same molecular formulas but only differ in their arrangement of atoms in space are called stereoisomers. More specifically, *cis* or *trans* isomers are called geometric isomers, a type of stereoisomer, because they only differ in their geometry. The following shows both *cis* and *trans* 1,2-dimethylcyclopentane.

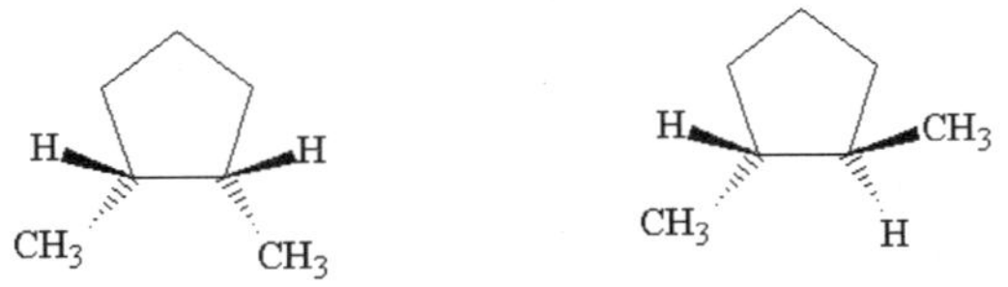

cis-1,2-dimethylcyclopentane   trans-1,2-dimethylcyclopentane

Even though cycloalkane rings are represented as planar structures when they are written, the actual structures of cycloalkane rings are puckered, which means the atoms are not in a single plane. The most stable structure of a ring is the one in which the C atoms bond to each other at an angle of 109.5°, the angle of sp^3 hybridized C atoms. As the angle deviates from this, angle strain (sometimes called Baeyer strain) is added to the molecule. For example, cyclopropane and cyclobutane have a total ring strain of 27.6 kcal (115 kJ) and 26.4 kcal (110 kJ), respectively. This is the result of the fact that the C atoms in these rings deviate greatly from the theoretical maximum angle of 109.5°. Cyclopentane only has a small amount of ring strain because its C atoms twist and can almost achieve the unstrained 109.5° bond angle.

Cyclohexane is a structure with the minimum possible ring strain because all of its C atoms have the 109.5° bond angle and all bonds are in a staggered conformation. The most stable conformation of cyclohexane is as follows.

Chair conformation
cyclohexane

This structure is called the chair conformation of cyclohexane. In this structure, each C atom is bonded to two C atoms and two H atoms, and all of the bond angles are 109.5°. Additionally, the torsional strain is minimized because all of the H atoms are staggered. When looking at the chair conformation of cyclohexane note that four C atoms are in one plane. One C atom is above that plane on one end of the molecule, and the other is below that plane on the other end of the molecule. The H atoms in this structure are classified as either axial or equatorial. The six axial H atoms are above and below the ring and parallel with the indicated axis of the ring. An axial H atom is found on every other C atom. The six equatorial H atoms extend out along the "equator" of the ring.

Axis

H ◄----------Axial H atom

H ◄------- Equatorial H atom

Another important conformation of cyclohexane is the boat conformation. It also has 109.5° bond angles, but it is less stable due to torsional strain because of the eclipsed H atoms and the van der Waals repulsions of the "flagpole" H atoms.

flagpole H atoms

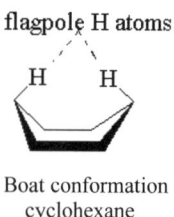

H   H

Boat conformation
cyclohexane

The boat conformation is 6.9 kcal (29 kJ) higher in energy than the chair.

When a substituent group is bonded to the cyclohexane ring, it is most stable when it is in an equatorial position. For example, studies show that 95% of methylcyclohexane is in the conformation in which the methyl group is in the equatorial position. Only 5% has the methyl group in the axial position. When the methyl group is in the equatorial position, it is farther from the adjacent H atoms than it is in the axial position, where it is repelled by the two axial H atoms on that side of the molecule.

1,3-diaxial interactions

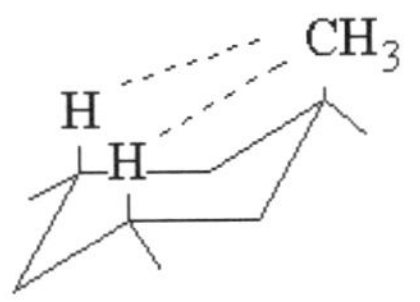

When two substituents are bonded to the cyclohexane ring, the groups can either be on the same side, *cis* to each other, or on opposite sides, *trans* to each other. Comparison of the enthalpy of combustion of the *cis* and *trans* isomers shows which isomer is more stable. For example, *trans*-1,2-dimethylcyclohexane is 1.5 kcal/mol (6.3 kJ/mol) more stable than *cis*-1,2-dimethylcyclohexane. How can this be explained? Consider the two possible *trans*-1,2-dimethylcyclohexane conformations.

(e,e) conformation (a,a) conformation

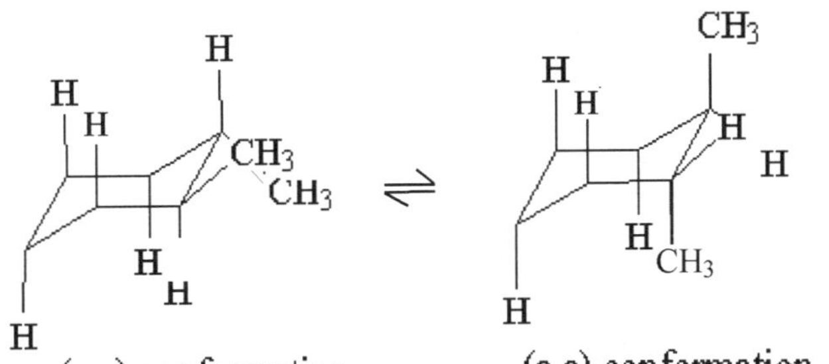

One of these structures has both methyl groups in the equatorial positions, (e,e), and the other has them in the axial positions, (a,a). The conformation in which the methyl groups are both in the equatorial positions is more stable because it minimizes the 1,3-diaxial interactions; thus, at room conditions nearly all of the *trans*-1,2-dimethylcyclohexane has the equatorial-equatorial conformation. In the *cis*-1,2-dimethylcyclohexane both conformations have a methyl group in an axial position; i.e., axial-equatorial conformation. This means that it will be less stable than the *trans* isomer because this structure will have 1,3-diaxial interactions with a methyl group in either conformation.

(a,e) conformation (a,e) conformation

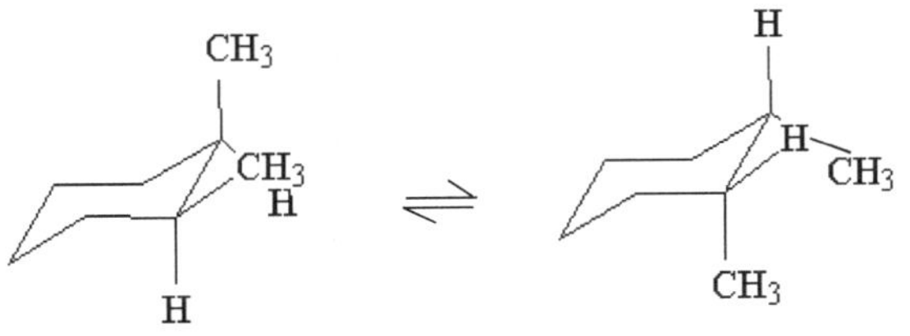

Cycloalkane rings can bond to each other. If the rings are joined by one C atom the molecule is classified as a spiro compound. An example of a spiro compound is spiro[2.4]heptane.

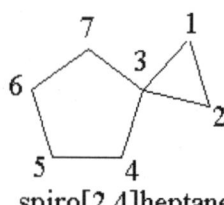

spiro[2.4]heptane

This compound has seven C atoms in three-membered and five-membered rings joined at the third C atom. Spiro compounds are rather rare. Fused cycloalkane rings are more common. In bicyclic compounds, ones with two fused rings, two common carbons are present. These two C atoms are called the bridgehead atoms. An example of a bicyclic system is norbornane.

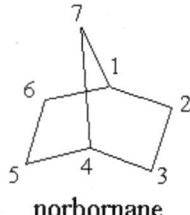

norbornane

The 1 and 4 C atoms are the bridgehead atoms in norbornane. The systematic name for norbornane is bicyclo[2.2.1]heptane. "Bicyclo" means that this compound has two fused rings. The "2.2.1" within brackets gives the number of C atoms bonded to the bridgehead atoms, and "heptane" states that seven C atoms are in the structure.

Test Yourself

1. What is the IUPAC name for the following molecule?

2. Draw the structure of 2-methyl-3-isopropylheptane.

3. Draw the structure of the isopropyl alkyl group.

4. What is the IUPAC name for the following molecule?

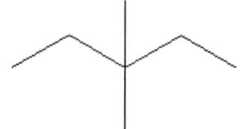

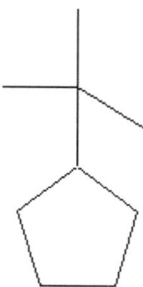

5. Draw the structures for all hexane isomers of heptane.

6. Write the names for all pentane isomers of heptane.

7. Which of the following compounds has the most negative enthalpy of combustion: isopentane, pentane, hexane, or heptane? Explain.

8. Draw the structures of three cycloalkane isomers with the formula C_5H_{10} each with a different size ring.

9. Write the common and IUPAC names for the following substituent group.

$$
\begin{array}{c}
CH_3 \\
| \\
CH_3 - C - CH_2 - \\
| \\
CH_3
\end{array}
$$

10. Draw the structure of dicyclopropylmethane showing all C and H atoms.

11. The normal boiling points for methane, ethane, and propane are $-160^\circ C$, $-88.7^\circ C$, and $-42.2^\circ C$, respectively. Explain this trend in alkane boiling points.

12. Draw the Lewis structure of *cis*-1,2-dimethylcyclohexane.

13. What is the relationship (same structure, constitutional isomers, geometric isomers, conformers, etc.) of structures I and II?

14. Draw the most stable conformation of t-butylcyclohexane.

15. Explain why the magnitude of the enthalpy of combustion of *trans*-1,2-dimethylcyclopropane is somewhat smaller than that of *cis*-1,2-dimethylcyclopropane.

16. Draw and explain the structure of bicyclo[3.1.1]heptane.

17. What are the possible locations (axial or equatorial) of the substituent groups in *cis*-1-methyl-3-isopropylcyclohexane that produce the most stable structure?

18. Write the balanced equation for the combustion of hexane.

19. Write the name of the following compound.

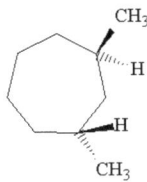

20. Which of the following has the highest boiling point?

    a. 2-methylpentane

    b. 2,3-dimethylbutane

    c. 2,2-dimethylbutane

21. Which of the following has the highest potential energy for pentane?

    a. gauche conformation

    b. totally eclipsed conformation

    c. anti conformation

    d. single eclipsed conformation

22. Which of the following is a correct IUPAC name for an alkane?

    a. 1-methyl-2-ethylhexane

    b. *cis*-2,3-dimethyloctane

    c. 2,3,4-trimethylheptane

    d. 3,4-ethyldecane

23. Which of the following alkyl groups would produce the largest 1,3-diaxial interactions in cyclohexane?

    a. ethyl

    b. propyl

    c. isopropyl

    d. t-butyl

24. Which would best describe the most stable conformation of *trans*-1-ethyl-3-methyl-cyclohexane?

    a. axial (methyl)-axial (ethyl)

    b. axial (methyl)-equatorial (ethyl)

    c. equatorial (methyl)-equatorial (ethyl)

    d. axial (ethyl)-equatorial (methyl)

25. Which of the following is a constitutional isomer of octane?

    a.   cyclooctane

    b.   3,3-dimethylheptane

    c.   4-isopropylheptane

    d.   tetramethylbutane

26. Which of the following releases the least energy when combusted?

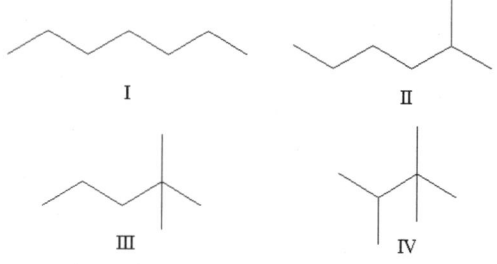

    a.   I

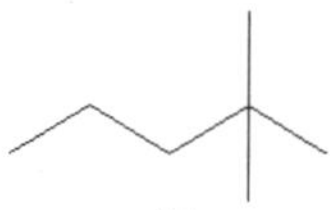

    b.   II

    c.   III

    d.   IV

27. What is the IUPAC name of the following?

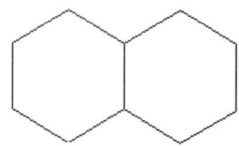

    a.   bicyclo[3.3.3]decane

    b.   bicyclo[3.3.0]octane

    c.   bicyclo[4.4.0]decane

    d.   bicyclo[4.3.1]nonane

28. Which of the following has the greatest angle strain?

    a.   methylcyclohexane

    b.   methylcyclopentane

    c.   methylcyclobutane

    d.   methylcyclopropane

29. Which of the following best describes the stability of the *cis* and *trans* isomers of 1,1,3,5-tetraethylcyclohexane?

    a.   The *cis* isomer is more stable than the *trans*.

    b.   The *trans* isomer is more stable than the *cis*.

    c.   Both have the same stability.

    d.   The stability of the *cis* and *trans* isomers cannot be determined with this information.

30. Consider the following equilibrium.

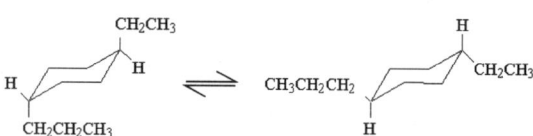

Which of the following best describes the equilibrium constant, K, for this interconversion?

    a. $K = 1$

    b. $K > 1$

    c. $K < 1$

    d. K cannot be predicted from this information.

✓ Check Yourself

1. The name of this compound is 3,3-dimethylpentane because two methyl groups are bonded to the third C atom in a chain of five C atoms. (**Alkane nomenclature**)

2.

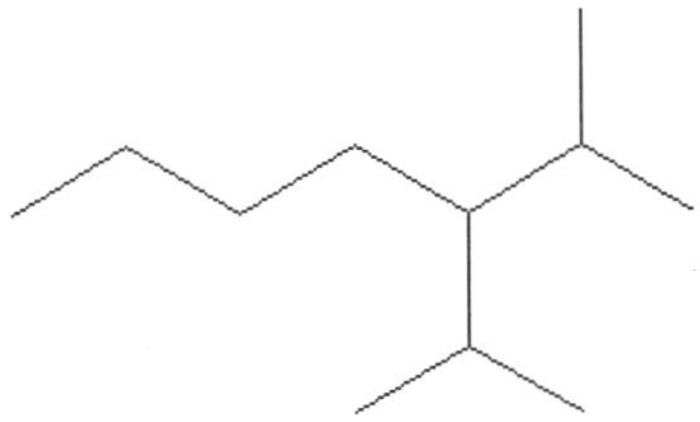

(Alkane structures)

3.

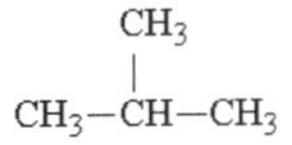

CH₃
|
CH₃—CH—CH₃

2-methylpropane
(isobutane)

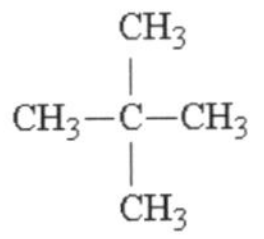

CH₃
|
CH₃—C—CH₃
|
CH₃

2,2-dimethylpropane
(neopentane)

(Alkyl groups)

4. The name of this compound is t-butylcyclopentane because a t-butyl group, $(CH_3)_3C$, is bonded to a five-membered saturated ring. (**Cycloalkane nomenclature**)

5.

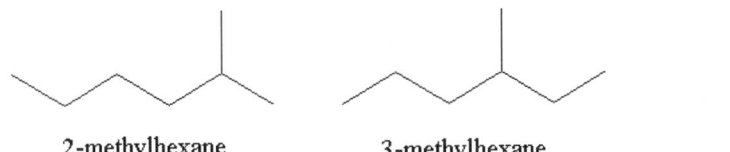

2-methylhexane   3-methylhexane

(Alkane isomers)

6. 2,2-dimethylpentane, 3,3-dimethylpentane, 2,3-dimethylpentane, 2,4-dimethylpentane, and 3-ethylpentane (**Alkane isomers**)

7. Heptane, C_7H_{16}, has the most negative enthalpy of combustion. As the number of C atoms in an alkane chain increases, the enthalpy of combustion becomes more negative, more exothermic. (**Enthalpy of combustion**)

8.

cyclopentane   methylcyclobutane   1,1-dimethylcyclopropane

Other correct molecules include ethylcyclopropane and 1,2-dimethylcyclopropane. (**Cycloalkane isomers**)

9. Its common name is a neopentyl group. Its IUPAC name is 2,2-dimethylpropyl group. (**Alkyl group nomenclature**)

10.

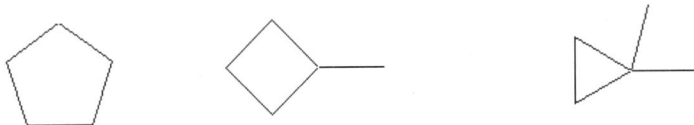

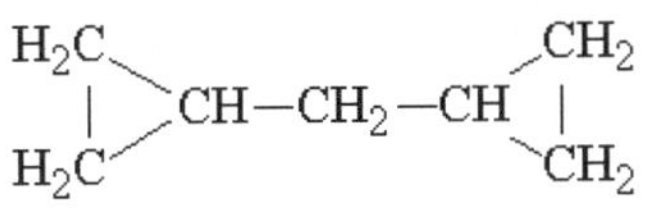

dicyclopropylmethane

(Cycloalkane structure)

11. Alkane molecules are nonpolar and interact via London intermolecular forces. Therefore, as the number of electrons increases with increasing size of the molecule (paralleling their molecular masses), the London forces increase, increasing the boiling point. (**Properties of alkanes**)

12.

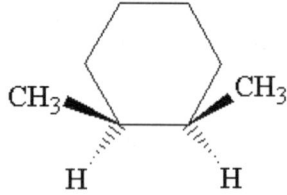

(Geometric isomers)

13. Structures I and II are the same gauche staggered conformations of the same molecule, 2-methylbutane, viewed from different directions. (**Conformational analysis**)

14.

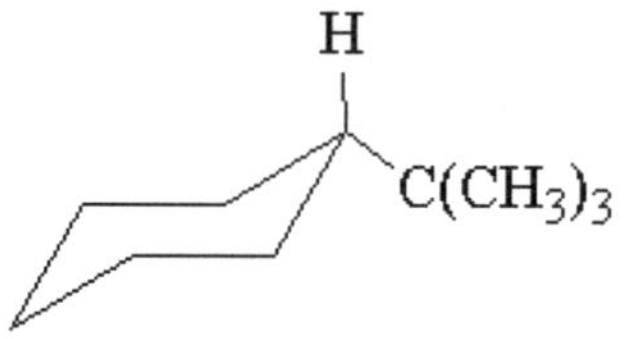

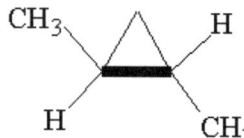

The most stable conformation places the bulky tert-butyl group in the equatorial position. Large groups such as tert-butyl are too large to occupy the axial position. (**Cycloalkane conformations**)

15. The smaller magnitude of the enthalpy of combustion for the *trans* isomers means that it is more stable. When the two methyl groups are on opposite sides of the cyclopropane ring, they are widely separated and do not interact. When these two groups are on the same side, they repel each other, making the molecule less stable. (**Geometric isomers**)

trans-1,2-dimethylcyclopropane 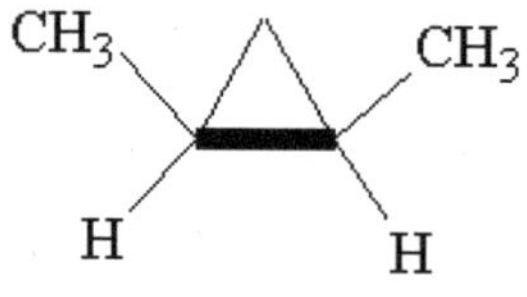   cis-1,2-dimethylcyclopropane

16.

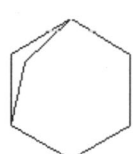

bicyclo[3.1.1]heptane

This compound has a total of seven C atoms. Bonded between the bridgehead atoms are one three-carbon chain and two one-carbon chains. (**Polycyclic alkanes**)

17. The most stable arrangement for *cis*-1-methyl-3-isopropylcyclohexane is the structure that has both groups in the equatorial positions, (e,e). The other structure, (a,a), is less stable due to diaxial interactions.

   (Cycloalkane structure)

18. $C_6H_{14} + 19/2O_2 \rightarrow 6CO_2 + 7H_2O$ **(Alkane reactions)**

19. The name of this compound is *trans*-1,3-dimethylcycloheptane. **(Cycloalkane nomenclature)**

20. a. 2-Methylpentane has the highest boiling point because it has only one branch in the chain. Alkane isomers with more branches have lower boiling points due to weaker London forces that result from less contact between molecules. **(Alkane properties)**

21. b. The totally eclipsed conformation has the highest energy because a methyl group on the second C atom is eclipsed with the ethyl group on the third C atom. **(Conformational analysis)**

22. c. 2,3,4-Trimethylheptane is the only correct IUPAC name. **(Alkane nomenclature)**

23. d. The *t*-butyl group is the bulkiest of those listed and would have the greatest van der Waals repulsions of the 1,3-diaxial H atoms. **(Cycloalkane conformations)**

24. b. The conformation with the methyl group in the axial position and the ethyl group in the equatorial position is most stable because the ethyl group is bulkier than the methyl group and would produce greater diaxial interactions. **(Cycloalkane conformations)**

25. d. Tetramethylbutane is the only compound that has the molecular formula of C_8H_{18}. **(Alkane isomers)**

26. d. IV is the most stable heptane isomer because it is most branched. The more highly branched isomers release less energy when combusted than the less highly branched ones. **(Enthalpy of combustion)**

27. c. Bicyclo[4.4.0]decane is the correct name for this compound because two four-carbon chains are attached between the bridgehead C atoms, which are bonded to each other. **(Polycyclic alkanes)**

28. d. Methylcyclopropane has a three-membered ring, which produces the greatest angle strain. **(Cycloalkane stability)**

29. a. The *cis* isomer is most stable because it can exist in a conformation in which only one ethyl group is in an axial position. The *trans* isomer has two ethyl groups in axial positions. **(Cycloalkane conformations)**

30. b. The value of *K* is greater than 1 because this equilibrium lies to the right. The substituent groups are most stable in the equatorial positions and would be less stable in the axial positions. **(Cycloalkane conformations)**

Grade Yourself

Circle the number of questions you missed, then fill in the total incorrect for each topic. If you answered more than three questions incorrectly, you need to focus on that topic. (If a topic has less than three questions and you had at least one wrong, we suggest you study that topic also. Read your textbook, a review book, or ask your teacher for help.)

Subject: Alkanes and Cycloalkanes

Topic	Question Numbers	Number Incorrect
Alkane nomenclature	1, 22	
Alkane structures	2	
Alkyl groups	3	
Alkane isomers	6, 25	
Enthalpy of combustion	7, 26	
Cycloalkane isomers	8	
Alkyl group nomenclature	9	
Cycloalkane structure	10, 17	
Properties of alkanes	11	
Geometric isomers	12, 15	
Conformational analysis	13, 21	
Cycloalkane conformations	14, 23, 24, 29, 30	
Polycyclic alkanes	16, 27	
Cycloalkane nomenclature	4, 19	
Alkane properties	20	
Cycloalkane stability	28	
Alkane reactions	18	

Functional Groups: Alcohols and Alkyl Halides

3

Brief Yourself

A functional group is an atom or group of atoms in an organic molecule that gives the molecule characteristic chemical and physical properties. The functional group usually is the site where bonds break and form in chemical reactions. Two important functional groups are the –OH group, found in alcohols, and the halogen atom, –X, in alkyl halides. Thus, the general formulas of alcohols and alkyl halides are ROH and RX, respectively.

To write the name of an alcohol, the *e* from the end of the corresponding alkane is removed and is replaced with the letters *ol*. For example the simplest alcohols are methanol and ethanol.

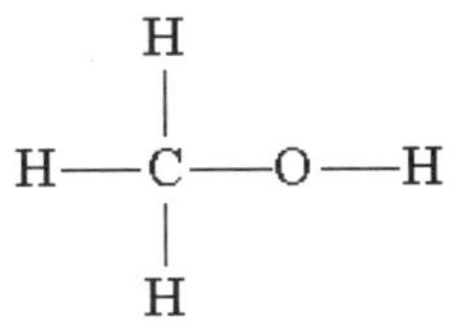

$$H-\overset{\displaystyle H}{\underset{\displaystyle H}{\overset{|}{\underset{|}{C}}}}-O-H \qquad H-\overset{\displaystyle H}{\underset{\displaystyle H}{\overset{|}{\underset{|}{C}}}}-\overset{\displaystyle H}{\underset{\displaystyle H}{\overset{|}{\underset{|}{C}}}}-O-H$$

methanol 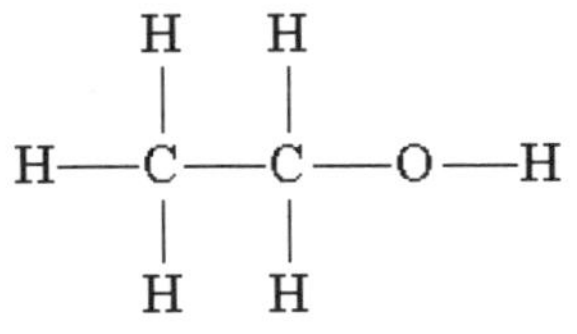   ethanol

Two constitutional isomers of propanol exist, 1-propanol and 2-propanol.

$$CH_3-CH_2-CH_2-OH \qquad CH_3-\overset{\displaystyle OH}{\overset{|}{CH}}-CH_3$$

1-propanol   2-propanol

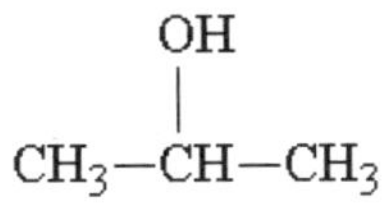

The common names of alcohols are generated by writing the name of the alkyl group in the molecule followed by the word "alcohol." Thus, the above structures are also called *n*-propyl alcohol and isopropyl alcohol.

The names of alkyl halides are derived in a similar manner to the names of alkanes except that the types and positions of the halogen atoms are identified. For example, consider the structure of 1,2-difluoroethane.

$$
\begin{array}{ccc}
 & H & H \\
 & | & | \\
F\!-\!\!\!\!& C\!-\!\!\!\!& C\!-\!F \\
 & | & | \\
 & H & H
\end{array}
$$

1,2-difluoroethane

If more than one halogen atom is in the molecule, they are listed alphabetically in the name. This is illustrated in 1-bromo-2-chlorohexane.

1-bromo-2-chlorohexane

Alcohols and alkyl halides are often classified as primary, secondary, or tertiary with respect to the location of the functional group. A primary alcohol ($1°$) or alkyl halide is a molecule that has a functional group bonded to a C atom (primary C atom) that is bonded only to one C atom. 1-Propanol is a primary alcohol and 1-chloropropane is a primary alkyl halide. A secondary alcohol ($2°$) or alkyl halide is one where the functional group is bonded to a C atom (secondary C atom) that is bonded to two C atoms. 2-Propanol is a secondary alcohol and 2-chloropropane is a secondary alkyl halide. A tertiary alcohol ($3°$) or alkyl halide is one where the functional group is bonded to a C atom (tertiary C atom) that is bonded to three C atoms. *t*-Butyl alcohol and *t*-butyl chloride are examples of molecules with tertiary C atoms.

$$
\begin{array}{c}
CH_3 \\
| \\
CH_3\!-\!C\!-\!CH_3 \\
| \\
OH
\end{array}
\qquad
\begin{array}{c}
CH_3 \\
| \\
CH_3\!-\!C\!-\!CH_3 \\
| \\
Cl
\end{array}
$$

t-butyl alcohol   t-butyl chloride
2-methyl-2-propanol   2-chloro-2-methylpropane

Alcohol molecules have the polar –OH group and thus have strong H-bonds as the forces between molecules, or intermolecular forces. As a result, alcohols have rather high melting and boiling points, and the low-molecular mass alcohols are water-soluble. Alkyl halides with similar molecular masses tend to have lower melting and boiling points than alcohols because of the weaker dipole-dipole interactions between their molecules. Alkyl halides are more nonpolar and are significantly less soluble in water than alcohols, but are more soluble in nonpolar organic solvents.

Alcohols can behave as Brønsted-Lowry acids or bases. A Brønsted-Lowry acid donates a proton, (H^+), and a Brønsted-Lowry base accepts a proton. When alcohols are in the presence of stronger bases, they behave as Brønsted-Lowry acids.

$$B^- + R\text{–}OH \rightleftarrows BH + R\text{–}O^-$$

In this equilibrium the strong base B^- removes a proton from the alcohol, ROH, producing the conjugate acid, BH, of B^- and an alkoxide ion, RO^-. Alkoxide ions are strong bases because they are the conjugate bases of alcohols, very weak acids.

The K_a values, acid-ionization equilibrium constants, for most alcohols are on the order of 10^{-16} to 10^{-18}. When alcohols are in the presence of stronger acids, they behave as Brønsted-Lowry bases.

$$ROH + HA \rightleftarrows ROH_2^+ + A^-$$

The alcohol accepts the proton from the acid and produces an alkyloxonium ion, ROH_2^+. An oxonium ion is one that has a positive charge on an O atom.

Alkyl halides can be produced from alcohols and hydrogen halides as follows.

$$ROH + HX \rightarrow RX + H_2O$$

A specific example of such a reaction occurs when *t*-butyl alcohol and hydrogen chloride react to produce *t*-butyl chloride and water.

$$(CH_3)_3C\text{–}OH + HCl \rightarrow (CH_3)_3C\text{–}Cl + H_2O$$

The order of reactivity of the hydrogen halides toward alcohols is as follows.

$$HI > HBr > HCl \gg HF$$

The order of reactivity of the alcohols reacting with hydrogen halides is as follows.

$$tertiary > secondary > primary > methyl$$

This means that elevated temperatures and higher concentrations are generally needed for reactions of methyl and primary alcohols than are needed for secondary and tertiary alcohols.

The reaction mechanism goes through a carbocation intermediate when a tertiary alcohol reacts with a hydrogen halide. The first step (a fast step) of the mechanism is protonation of the alcohol by the acid. Next, the C–O bond breaks, producing a carbocation and water. This second step of the mechanism is the rate-determining step (also called rate-limiting step), the slowest step. The final step (a fast step) is the attack on the carbocation by the halide ion, producing the alkyl halide.

A carbocation is an ion that has a positive charge on a C atom. Carbocations can be methyl (CH_3^+), primary (1°), secondary (2°), or tertiary (3°). Consider the following primary, secondary, and tertiary carbocations.

$$\begin{array}{ccc}
\overset{\displaystyle H \quad H}{\underset{\displaystyle H \quad H}{H-C-C^+}} & \overset{\displaystyle H \quad H \quad H}{\underset{\displaystyle H \quad H}{H-C-\underset{+}{C}-C-H}} & \underset{\displaystyle CH_3}{CH_3-\overset{+}{C}-CH_3} \\
\text{ethyl carbocation} & \text{isopropyl carbocation} & \text{t-butyl carbocation} \\
\text{(primary)} & \text{(secondary)} & \text{(tertiary)}
\end{array}$$

The carbocation C atom is sp^2 hybridized; thus, its structure is trigonal planar with 120° bond angles. The stability of carbocations is as follows.

$$3° > 2° > 1° > CH_3^+$$

Decreasing carbocation stability

This order is explained in terms of the electron-releasing ability of alkyl groups.

Tertiary carbocations are more stable than secondary carbocations because three R groups donate more electron density into the positive C atom than two R groups. Positively charged C atoms are stabilized whenever they are bonded to electron-releasing groups and are destabilized when bonded to electron-withdrawing groups.

As a result of the positive charges on carbocations, they are strong electrophiles (species that accept electrons). Electrophiles are attacked by nucleophiles (species that donate electrons). In the reaction of alcohols and hydrogen halides to form alkyl halides, the nucleophile that attacks the carbocation is the halide ion, X^-. This occurs in the last step of the mechanism.

A different mechanism is followed when a halide ion attacks methyl alcohol or a primary alcohol. Initially, the acid protonates the alcohol producing an alkyloxonium ion. At this point the mechanism is concerted, a mechanism that takes place in one step. The halide ion attacks from the opposite side of the OH_2^+ group as follows.

$$X^- \;+\; \underset{\substack{\text{primary} \\ \text{alkyloxonium ion}}}{\overset{\displaystyle R}{\underset{\displaystyle H}{H-C-\overset{+}{O}H_2}}} \;-\; \left[\underset{\substack{\text{activated complex}}}{\overset{\displaystyle R}{\underset{\displaystyle H \quad H}{X^{\delta-}\cdots C\cdots OH_2^{\delta+}}}} \right] \;-\; \underset{\substack{\text{primary alkyl halide}}}{\overset{\displaystyle R}{\underset{\displaystyle H}{X-C-H}}} + H_2O$$

Alkyl halides can also be produced by other reagents. Alcohols react with thionyl chloride, $SOCl_2$, in pyridine and produce alkyl chlorides.

$$ROH + SOCl_2 \rightleftarrows RCl + SO_2 + HCl$$

Alcohols also react with phosphorus tribromide, PBr_3, and produce alkyl bromides, RBr.

$$3ROH + PBr_3 \rightleftarrows 3RBr + P(OH)_3$$

Alkyl halides are produced from alkanes. For example, if methane is mixed with chlorine gas in the presence of light, $h\upsilon$, or at an elevated temperature, methyl chloride (chloromethane) and hydrogen chloride are the products.

$$CH_4 + Cl_2 \rightleftarrows CH_3Cl + HCl$$

If excess Cl_2 is present, this reaction will continue producing methylene chloride (dichloromethane), chloroform (trichloromethane), and carbon tetrachloride (tetrachloromethane).

$$CH_3Cl + Cl_2 \rightleftarrows CH_2Cl_2 + HCl$$

$$CH_2Cl_2 + Cl_2 \rightleftarrows CHCl_3 + HCl$$

$$CHCl_3 + Cl_2 \rightleftarrows CCl_4 + HCl$$

The high-energy species produced in the chlorination of methane are free radicals. A free radical is a neutral species that has an unpaired electron. Free radicals can be methyl, primary, secondary, or tertiary. The order of stability of free radicals is as follows.

$$3^o > 2^o > 1^o > CH_3$$
Decreasing free radical stability

The chlorination of methane follows a free-radical chain mechanism. The first step, the chain-initiating step, is the cleavage of the Cl–Cl bond by either light or heat. This produces two Cl free radicals, Cl·. The next two steps are the chain-propagating steps. The Cl free radical attacks and breaks a C–H bond, producing HCl and a methyl free radical, CH_3·.This reactive methyl free radical then attacks a Cl_2 molecule, producing methyl chloride, a product, and another Cl free radical that can continue the chain propagation by attacking another methane molecule. The chain-terminating steps occur when two free radicals combine; e.g., two Cl free radicals, a Cl and methyl free radical, or two methyl free radicals.

Chlorine in the presence of light or heat attacks any alkane, producing one or more alkyl halides. When Cl_2 reacts with butane under the proper conditions, two products are obtained, 1-chlorobutane and 2-chlorobutane. At 35°C, with light, the product mixture has 28% 1-chlorobutane and 72% 2-chlorobutane. More of the 2-chlorobutane is obtained because the mechanism to produce it follows a lower energy pathway through the more stable secondary free radical than the higher energy pathway through the primary free radical.

Bromination of alkanes is similar to chlorination. However, if Br_2 reacts with butane in the presence of light or heat, the principal product is 2-bromobutane with very little 1-bromobutane. This means that bromination is more selective than chlorination. The reason bromination is more selective than chlorination is that the transition state during bromination resembles an alkyl free radical (a late-forming transition state) more than does the transition state (an early-forming transition state) in chlorination. The transition state in chlorination resembles the alkane more. This difference is a result of the differences in energies of removing the H atom in bromination and chlorination.

 **Test Yourself**

1. Write the IUPAC and common name for the following alcohol.

2. Write the IUPAC name for the following alcohol.

3. Draw the structure of 1,5-dibromo-2-methyl-2-heptanol and classify it as a primary, secondary, or tertiary alcohol.

4. Draw the structure of *cis*-1-iodo-3-fluorocyclohexane.

5. Write the name for the following alkyl halide.

6. Draw the Newman projection for the anticonformation of 1,2-dichloroethane.

7. Draw the structures and write the names for all constitutional isomers of dibromopropane.

8. Write an equation that shows how 2-methyl-2-hexanol acts as a Brønsted-Lowry base.

9. Draw the structure of the product of the reaction of 1-methylcyclohexanol and HCl.

10. What is the product of the reaction of potassium and *t*-butyl alcohol?

11. a. What is the principal type of reaction intermediate that forms when an alkyl halide is produced from a tertiary alcohol and hydrogen halide?

    b. What is the principal type of reaction intermediate that forms in the halogenation of alkanes?

12. a. Draw the structure of the carbocation intermediate that forms in the reaction of 1-methylcyclobutanol and hydrogen chloride.

    b. What type of carbocation is this?

    c. What is the geometry at the carbocation C atom?

13. Draw the structure of the most stable carbocation with the formula $C_5H_{11}^+$.

14. What is the product of the reaction of cycloheptanol and phosphorus tribromide?

15. Explain the most important features of the mechanism of the free radical halogenation of methane.

16. Write the chain-propagating steps for the chlorination of ethane.

17. Draw the structure of the transition state when a Cl free radical abstracts (removes) a H atom from the second C atom in propane. Describe this structure.

18. Draw the structure of the highest-energy transition state in the reaction of t-butyl alcohol with hydrogen chloride.

19. Write the names of all of the monochloride products of the free-radical chlorination of methylcyclobutane.

20. Which of the following compounds yields four monochloride products when it undergoes free radical chlorination?

    a. 2,2-dimethylpropane

    b. pentane

    c. 2,2-dimethylbutane

    d. 2-methylbutane

21. Which of the following reagents could be used to convert cyclohexanol to chlorocyclohexane?

    a. $SOCl_2$

    b. PBr_3

    c. Cl_2, light

    d. none of these

22. Classify the following carbocation.
    $$CH_3CH_2CH_2CH_2^+$$

    a. methyl

    b. $1°$

    c. $2°$

    d. $3°$

23. Consider the following carbocations.

 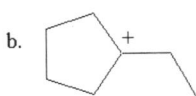

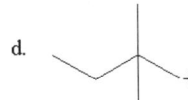

Which is most stable carbocation?

24. Which of the following is most selective when it reacts with alkanes in the presence of light?

    a. F_2

    b. Cl_2

    c. Br_2

    d. H_2

25. Which of the following is a correct statement regarding the effect of a Cl atom bonded to the C atom adjacent to a carbocation C?

    a   It helps stabilize the carbocation.

    b. It helps destabilize the carbocation.

    c. It has no effect on the stability of the carbocation.

    d. none of these

26. What effect does doubling the concentration of HCl have on the rate of the reaction of t-butyl alcohol and HCl?

    a. It doubles the rate of the reactions.

    b. It decreases the rate to one-half.

    c. It quadruples the rate of the reaction.

    d. It has no effect on the reaction rate.

27. What is the IUPAC name for the following?

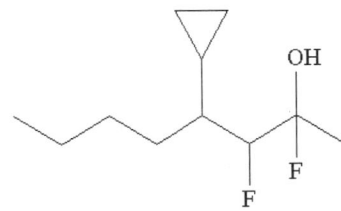

    a. 1,2-difluoro-3-cyclopropaneheptanol

    b. 1-[1,2-difluoro-2-hydroxybutyl]-1-butylcyclopropane

    c. 3-cyclopropyl-1,2-difluoro-1-methyl-1-heptanol

    d. 4-cyclopropyl-2,3-difluoro-2-octanol

28. Which of the following is the major product of the following reaction?

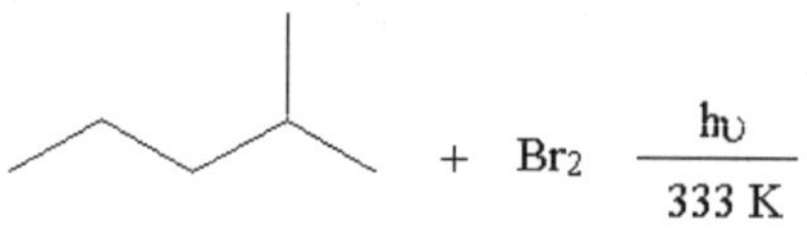

 + Br₂ $\xrightarrow[333\ K]{h\upsilon}$

    a. 1-bromo-2-methylpentane

    b. 2-bromo-2-methylpentane

    c. 3-bromo-2-methylpentane

    d. 4-bromo-2-methylpentane

    e. 1-bromo-4-methylpentane

29. An unknown compound has a molecular mass of 84 and has only C and H atoms. When it undergoes chlorination in the presence of light, three monochlorinated products are isolated. Identify this compound.

    a. hexane

    b. cyclohexane

    c. methylcyclopentane

    d. 1,3-dimethylcyclobutane

30. Which of the following is **not** true about transition states of organic reactions?

    a. The highest-energy transition state in a re-action mechanism is in the rate-determining step.

    b. Factors that stabilize the highest-energy transition state tend to increase the rate of the reaction.

    c. The activation energy must be added to a reaction so that it can reach the transition state and produce the activated complex.

    d. At higher temperatures, more reactant particles have energies that equal or exceed the energy associated with the transition state than at lower temperatures.

    e. All of the above are correct.

✔ Check Yourself

1. Its IUPAC name is 2,2-dimethyl-1-propanol, and its common name is neopentyl alcohol. (**Alcohol nomenclature**)

2. *cis*-1-methyl-2-propylcyclopentanol (**Alcohol nomenclature**)

3.

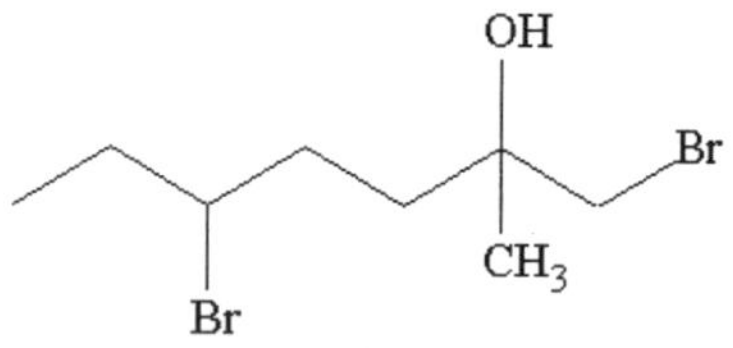

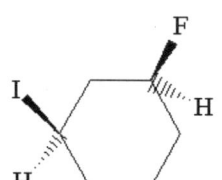

    This is a tertiary alcohol because the OH group is bonded to a C atom that is bonded to three C atoms. (**Alcohol structure**)

4.

(**Alkyl halide structure**)

5. 1,2,4-trichloro-3,5-difluorooctane (**Alkyl halide structure**)

6.

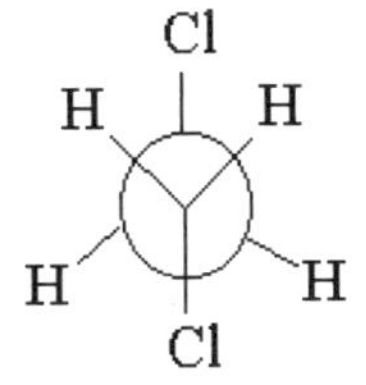

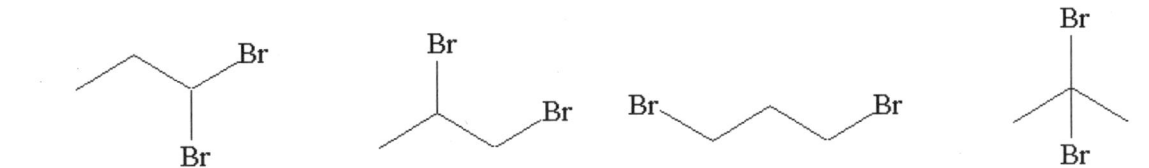

(Alkyl halide conformation)

7.

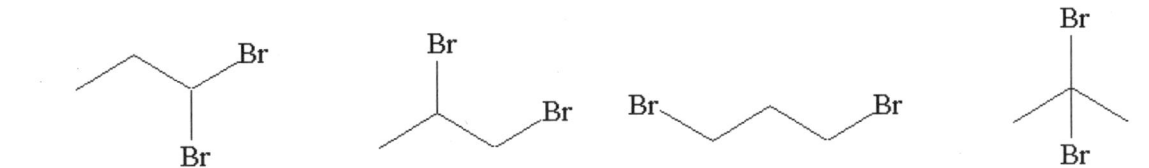

1,1-dibromopropane   1,2-dibromopropane   1,3-dibromopropane   2,2-dibromopropane

(Alkyl halide isomers)

8.

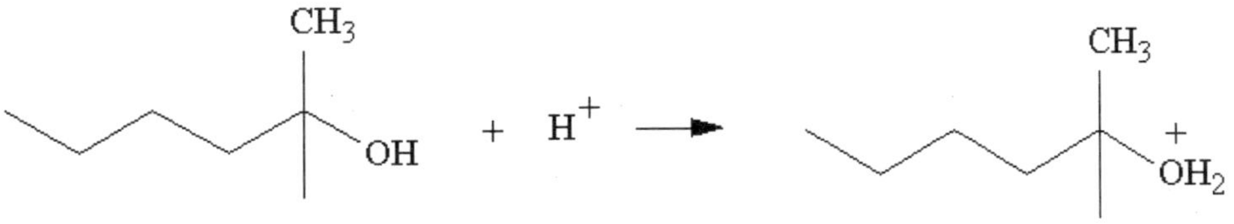

(Alcohol acid-base behavior)

9.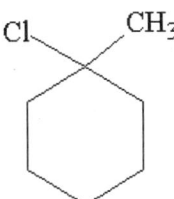

(Alcohol reactions)

10. Potassium can remove the H from the OH group in t-butyl alcohol, producing potassium t-butoxide, $K^+(CH_3)_3CO^-$. (**Alcohol reactions**)

11. a. carbocation intermediate,
    b. free radical intermediate (**Alcohol and alkane reactions**)

12.

a.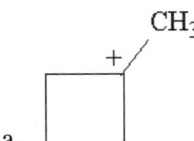

b. This is a tertiary carbocation.
c. It is trigonal planar. (**Carbocations**)

13.

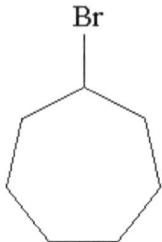

$$CH_3-CH_2-\underset{\underset{CH_3}{|}}{\overset{\overset{CH_3}{|}}{C}}+$$

This is the only isomer that produces a tertiary carbocation. **(Carbocation stability)**

14.

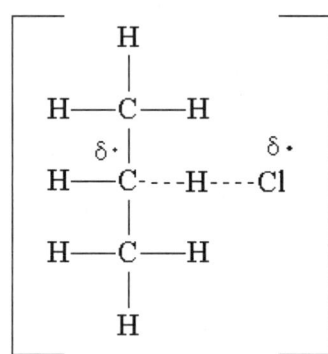

bromocycloheptane **(Alcohol reactions)**

15. The first step is the chain-initiating step in which the Cl–Cl bond is broken. The second set of major steps is the chain-propagating steps in which the methyl and Cl free radicals are produced repeatedly. Finally, the reaction is stopped by the chain-terminating steps in which free radicals bond. **(Free radical halogenation)**

16. 1. $CH_3CH_3 + Cl^{\cdot} \rightleftarrows CH_3CH_2^{\cdot} + HCl$

   2. $CH_3CH_2^{\cdot} + Cl_2 \rightleftarrows CH_3CH_2Cl + Cl^{\cdot}$ **(Free radical halogenation)**

17.

$$\left[\begin{array}{c} \overset{\overset{H}{|}}{H-C-H} \\ \overset{\delta\cdot|}{H-C----H----Cl} \quad \delta\cdot \\ \underset{\underset{H}{|}}{H-C-H} \end{array}\right]$$

The transition state shows that the C–H bond is partially broken and the C–Cl bond is partially formed. The Cl atom is losing its free-radical character and it is being picked up by the C atom. **(Free radical halogenation)**

18.

$$\left[\begin{array}{c} CH_3 \\ \overset{\delta+}{|} \quad \overset{\delta+}{} \\ H_3C\!-\!\!-\!C\!-\!-\!-\!-\!OH_2 \\ | \\ CH_3 \end{array}\right]$$

The C–O bond breaks in the alkyloxonium ion in the transition state, producing the *t*-butyl carbocation **(Alcohol reactions)**

19. The products of the reaction are: chlorocyclobutylmethane, 1-chloro-1-methylcyclobutane, 2-chloro-1-methylcyclobutane (*cis* and *trans*), and 3-chloro-1-methylcyclobutane (*cis* and *trans*). **(Halogenation of alkanes)**

20. d. 2-Methylbutane produces 1-chloro-2-methylbutane, 2-chloro-2-methylbutane, 2-chloro-3-methylbutane, and 1-chloro-3-methylbutane. **(Halogenation of alkanes)**

21. a. $SOCl_2$ is the only listed reagent that will effect this conversion. **(Alcohol reactions)**

22. b. It is a primary carbocation, $1°$, because the positive charge is located on a terminal C atom. **(Carbocations)**

23. b. This is the only tertiary carbocation; thus, it is the most stable of the four ions. **(Carbocations)**

24. c. Br_2 is the most selective of the three in free-radical halogenation of alkanes. **(Free radical halogenation)**

25. b. It will make the carbocation less stable because its high electronegativity withdraws electron density and makes the carbocation more positive. Increasing charge destabilizes ions. **(Carbocation stability)**

26. d. It has no effect on the reaction rate because HCl is not involved in the rate-determining step of this reaction (the formation of the carbocation). **(Alcohol reactions)**

27. d. 4-Cyclopropyl-2,3-difluoro-2-octanol is the correct name because the longest chain has eight C atoms. Bonded to the fourth C atom is a cyclopropyl group. Bonded to the second and third C atoms are F atoms. Finally, the OH group is bonded to the second C atom of the longest chain. **(Alcohol nomenclature)**

28. b. 2-Bromo-2-methylpentane is the product that results when the Br atom bonds to the only tertiary C atom. Br is very selective and a significant percentage of the reaction mixture is this product. **(Free radical halogenation)**

29. d. 1,3-Dimethylcyclobutane has the molecular formula of C_6H_{12}, which has a molecular mass of 84. When chlorinated the Cl atom can bond at the 1 and 2 positions on the ring and the methyl group; thus, three monochlorinated products could form. Cyclohexane only forms one monochlorinated product, and methylcyclopentane can form four products. **(Alkane reactions)**

30. e. All of the above are correct.

Grade Yourself

Circle the number of questions you missed, then fill in the total incorrect for each topic. If you answered more than three questions incorrectly, you need to focus on that topic. (If a topic has less than three questions and you had at least one wrong, we suggest you study that topic also. Read your textbook, a review book, or ask your teacher for help.)

Subject: Functional Groups: Alcohols and Alkyl Halides

Topic	Question Numbers	Number Incorrect
Alcohol nomenclature	1, 2, 27	
Alcohol structure	3	
Alkyl halide structure	4, 5	
Alkyl halide conformation	6	
Alkyl halide isomers	7	
Alcohol acid-base behavior	8	
Alcohol reactions	9, 10, 14, 18, 21, 26	
Alcohol and alkane reactions	11, 29	
Carbocations	12, 22, 23	
Carbocation stability	13, 25	
Free radical halogenation	15, 16, 17, 24, 28	
Halogenation of alkanes	19, 20	

Alkenes— Structure, Stability, and Preparations

4

Brief Yourself

Alkenes are hydrocarbons that have a C–C double bond. If two double bonds are in a molecule, it is called an alkadiene. If it has three double bonds then it is an alkatriene. The IUPAC name of an alkene is obtained by removing the *ane* from the end of the corresponding alkane and replacing it with *ene*. To designate the location of the double bond, the chain is numbered so that the double bond has the lowest possible value that corresponds to the number of the first C atom of the double bond. Consider the following examples: 1-pentene and 2-pentene.

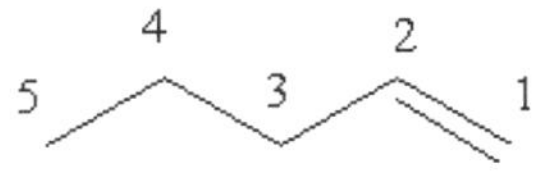

1-pentene 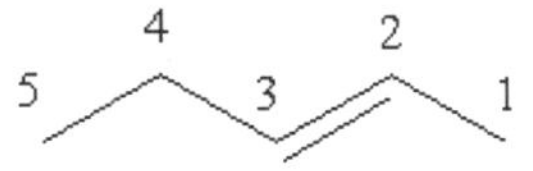   2-pentene

Substituent groups are identified in alkenes the same way they are in alkanes. For example, consider the following substituted alkene, 2,6-dimethyl-3-ethyl-3-heptene.

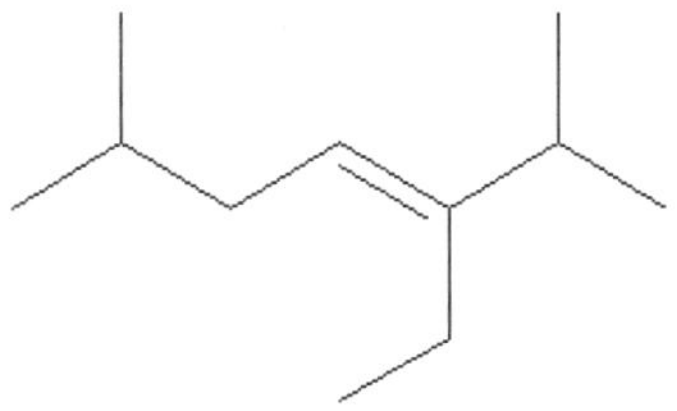

2,6-dimethyl-3-ethyl-3-heptene

Cycloalkenes are named similarly to cycloalkanes. In all cycloalkenes the double bond is always between the first and second C atom; thus, there is no need to identify its position. The location of substituent groups on a cycloalkene ring are relative to the position of the double bond. Consider the following examples of cycloalkenes.

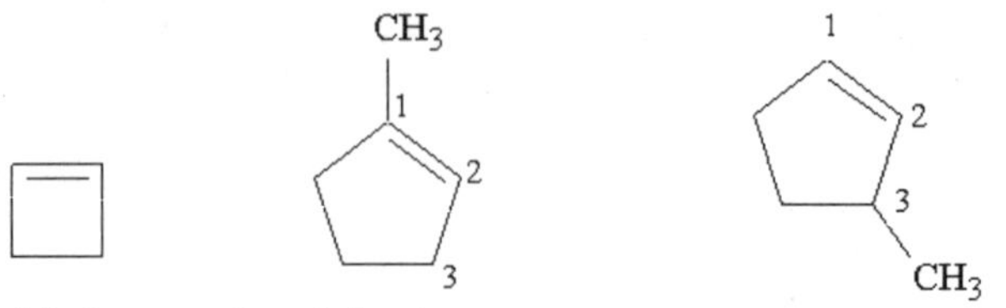

cyclobutene   1-methylcyclopentene   3-methylcyclopentene

Sometimes a C–C double bond is a substituent group bonded to a functional group or ring. The vinyl and allyl groups are most commonly encountered.

$$CH_2=CH- = \text{vinyl group } (CH_2=CH-Cl = \text{vinyl chloride})$$

$$CH_2=CHCH_2- = \text{allyl group } (CH_2=CHCH_2-Br = \text{allyl bromide})$$

Common names of alkenes are obtained by adding the ending *ylene* to the name of the C skeleton. For example, the two simplest alkenes are ethylene (ethene) and propylene (propene).

The physical properties of alkenes resemble those of the alkanes because both groups are nonpolar with London forces between the molecules. The lower-molecular-mass alkenes are gases. As the molecular mass increases, the London forces increase and the alkenes become liquids and solids. Alkenes are most soluble in nonpolar solvents.

The structure of the C–C double bond determines most of the properties of alkenes. The C–C double bond consists of a strong sigma bond, σ, that results from the overlap of an sp^2 orbital from one C atom with an sp^2 orbital from the other. This leaves the two unhybridized p orbitals, which are located perpendicular to the plane of the sp^2 hybrid orbitals. These p orbitals overlap sideways and produce the pi bond, π, of the double bond.

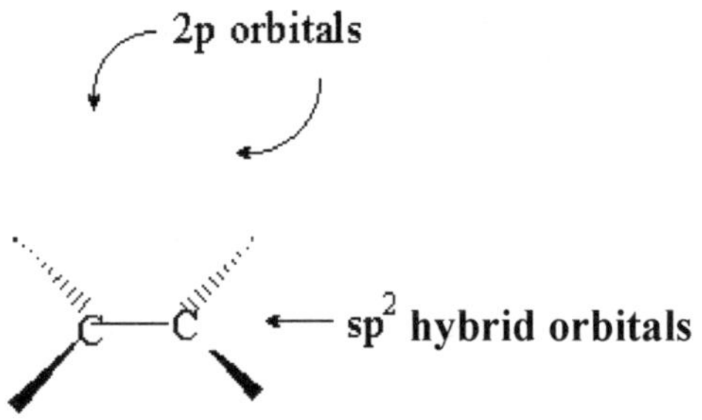

The σ bond is a strong bond that is difficult to break. The π bond is weaker and readily breaks when attacked by electrophiles, species that accept electrons. The π bond is strong enough to prevent rotation around the double bond at room conditions.

Because of the restricted rotation around the double bond many alkenes exist as stereoisomers, more specifically as geometric isomers. If the two R groups bonded are on the same side of the double bond, they are *cis* isomers, and if they are on opposite sides, they are *trans* isomers. The following shows the structure of *cis*-2-pentene and *trans*-2-pentene.

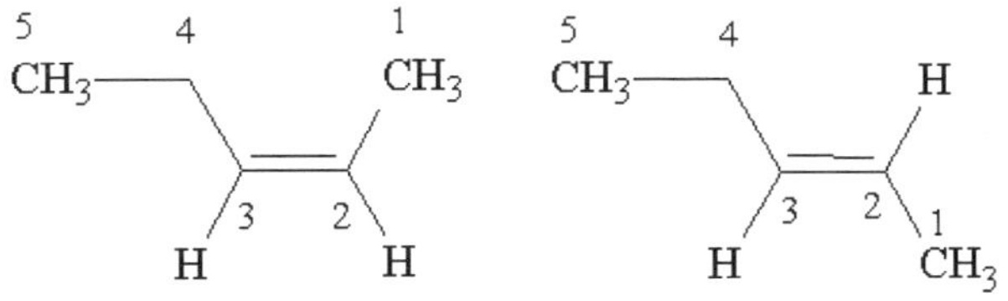

cis-2-pentene   trans-2-pentene

In a *cis*-alkene the longest continuous chain starts on one side of the double bond and ends on the same side. In a *trans*-alkene the chain starts on one side and ends on the other.

For complicated alkenes it is more difficult to determine if the structure is *cis* or *trans*. For example, consider the following halogen-substituted alkene.

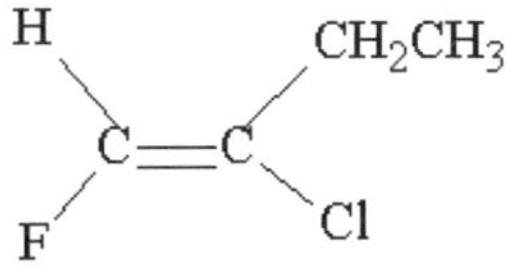

Is this *cis* or *trans* 2-chloro-1-fluoro-1-butene? To answer this question the IUPAC system uses the *E-Z* system to easily designate *cis* and *trans* isomers. *E* is usually associated with *trans*-alkenes and *Z* with *cis*-alkenes. To decide if this compound is the *E* or *Z* isomer, determine which of the two groups bonded to each C atom in the double bond has a higher priority. Priority depends on atomic number. An atom with a higher atomic number has a higher priority. In this molecule, F has a higher priority than H and Cl has a higher priority than C. Therefore, this molecule is (Z)-2-chloro-1-fluoro-1-butene.

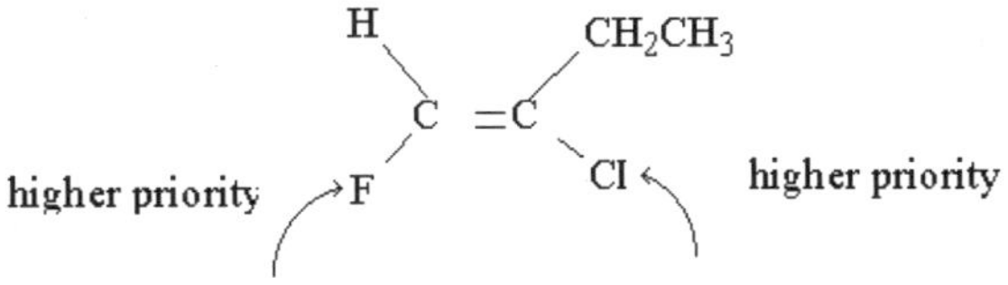

higher priority   higher priority

(Z)-2-chloro-1-fluoro-1-butene

If the priority cannot be determined from the atoms bonded directly to the double bond, then the atoms bonded to these atoms are examined until the first place where they differ. For example, is the following the *E* or *Z* isomer?

higher priority

(H, H, Cl)   ClCH$_2$CH$_2$ CH$_2$CH$_3$ (H, H, C)

C =C

(H, H, C) CH$_3$CH$_2$CH$_2$ CH$_2$OH (H, H, O)

higher priority

This is an *E*-alkene because the two higher priority groups are on opposite sides of the alkene. Each C atom bonded to the double bond is a C atom bonded to two H atoms plus another. These atoms are used to rank the groups bonded to the double bond.

The stability of alkenes may be determined from their enthalpies of hydrogenation, -$\Delta H_{hydrogenation}$, which is the amount of heat released when an alkene is converted to an alkane by catalytic addition of H$_2$. For example, 32.8 kcal/mol (137 kJ/mol) is released when ethene is hydrogenated on a Pt catalyst, and only 30.1 kcal/mol (126 kJ/mol) is released when propene is hydrogenated.

$$H_2C=CH_2 + H_2 \rightarrow H_3CCH_3 \qquad\qquad \Delta H = -32.8 \text{ kcal/mol}$$

$$CH_3CH=CH_2 + H_2 \rightarrow H_3CCH_2CH_3 \qquad \Delta H = -30.1 \text{ kcal/mol}$$

Because propene releases less energy on hydrogenation, it is more stable than ethene. In general, more highly substituted alkenes are more stable and thus have lower enthalpies of hydrogenation than those that are less substituted.

H   R   H   R   H   R   R   R

C =C   <   C =C   <   C =C   <   C =C

H   H   R   H   R   R   R   R

Monosubstituted   Disubstituted   Trisubstituted   Tetrasubstituted

Increasing alkene stability ⟶

Alkenes can be prepared by β-elimination reactions (also called 1,2-elimination reactions). A β-elimination occurs when two atoms or groups on adjacent C atoms are removed from a molecule. Dehydration of alcohols is an example of a β-elimination alkene preparation reaction. If an alcohol is heated in the presence of a dehydrating acid such as sulfuric acid, H$_2$SO$_4$, or phosphoric acid, H$_3$PO$_4$, a H atom from one C and the OH group from an adjacent C atom are removed. An example of this reaction is the dehydration of ethanol to produce ethene (ethylene).

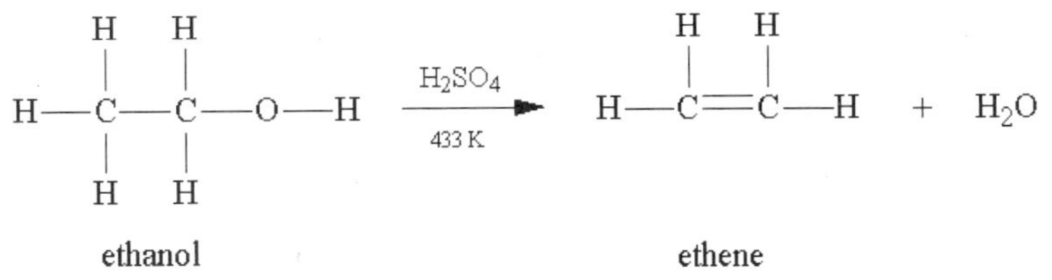

ethanol   ethene

The acid is a catalyst in this reaction.

The acid-catalyzed dehydration of alcohols follows Saytzeff's rule, which states that the most highly substituted alkene is the major product. For example, 1-methylcyclopentene is the major product of the following reaction.

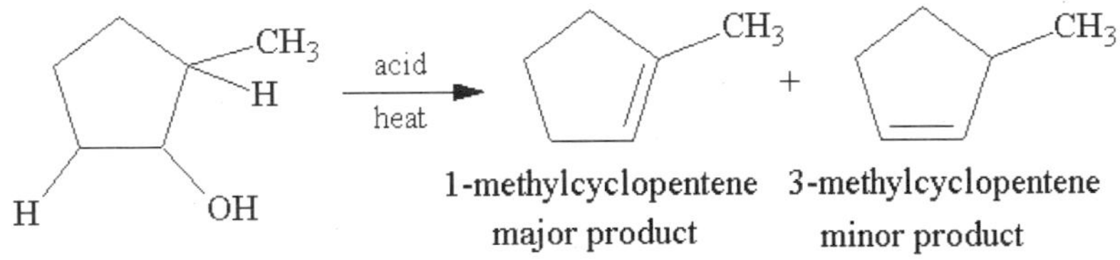

1-methylcyclopentene 3-methylcyclopentene
major product   minor product

1-Methylcyclopentene is a trisubstituted alkene and 3-methylcyclopentene is only disubstituted. Saytzeff's rule can best be explained in terms of the stability of the alkene products. The more highly substituted alkene is more stable than the less substituted one; thus, the most stable alkene is the product of the β-elimination of alcohols. This reaction is classified as a regioselective reaction because one constitutional isomer is produced in greater amounts than another. This reaction is also stereoselective because the *trans*-isomer is usually found in greater amount than the *cis*-isomer.

The mechanism for the dehydration of alcohols follows three elementary steps. First, the acid protonates the alcohol and produces an alkyloxonium ion, ROH_2^+. Next, the C–O bond breaks in the alkyloxonium ion and a carbocation results. Finally, the conjugate base of the acid removes a proton (H^+) from the carbocation, producing the alkene. The rate-determining step is the formation of the carbocation; hence, tertiary alcohols dehydrate more rapidly than secondary alcohols, which dehydrate more readily than primary alcohols.

As a result of the formation of a carbocation, rearrangements occur in some dehydration reactions. A rearrangement occurs when one or more of the products of a reaction has a different C skeleton than the reactant. A rearrangement can also occur when a H atom migrates to another C atom.

A rearrangement occurs when 3,3-dimethyl-2-butanol is dehydrated using phosphoric acid and heat.

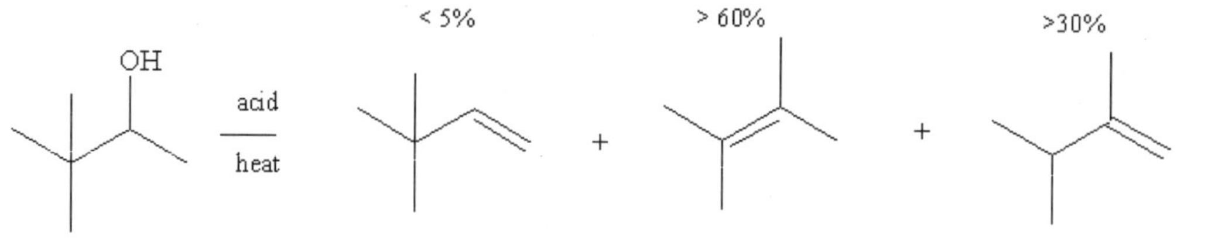

< 5%	> 60%	>30%

3,3-dimethyl-2-butanol   3,3-dimethyl-1-butene   2,3-dimethyl-2-butene   2,3-dimethyl-1-butene

The two rearranged products are the major products, and the one with the same C skeleton is formed in much smaller amounts. The rearrangement that occurs in this reaction is a methyl shift (also called a methyl migration) to produce a more stable tertiary carbocation from a secondary carbocation. This 1,2-methyl shift (migration of a group from one C atom to the next) occurs as follows.

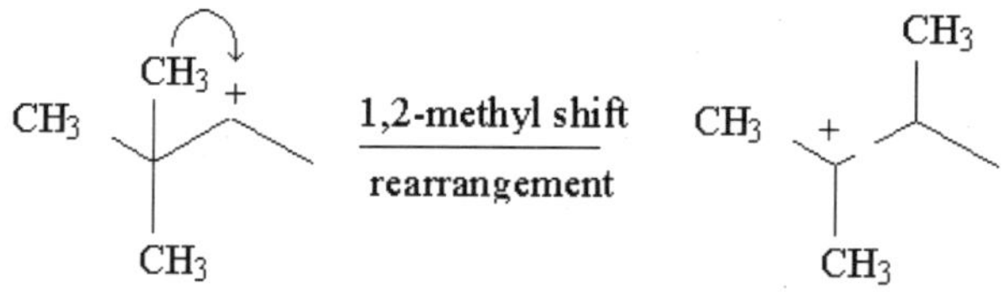

secondary carbocation   tertiary carbocation

Besides alkyl groups, H atoms can also migrate to another C atom in a carbocation mechanism. This is a hydride shift.

Another way to prepare alkenes is through dehydrohalogenation, the β-elimination of a H and halogen, X, on adjacent C atoms.

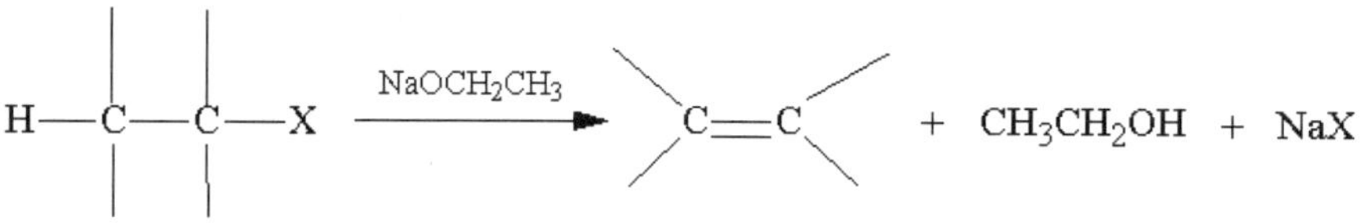

alkyl halide   alkene

This reaction requires an alkyl halide and a strong base such as sodium ethoxide, $NaOCH_2CH_3$, or potassium hydroxide, KOH. The products of the reaction are an alkene, the conjugate acid of the base, and a salt.

The mechanism for dehydrohalogenation is either E2 or E1. E2 means elimination bimolecular, and E1 means elimination unimolecular. In the E2 mechanism, C–X bond breaking, C–C double bond formation, and C–H bond breaking all occur simultaneously. It occurs as follows:

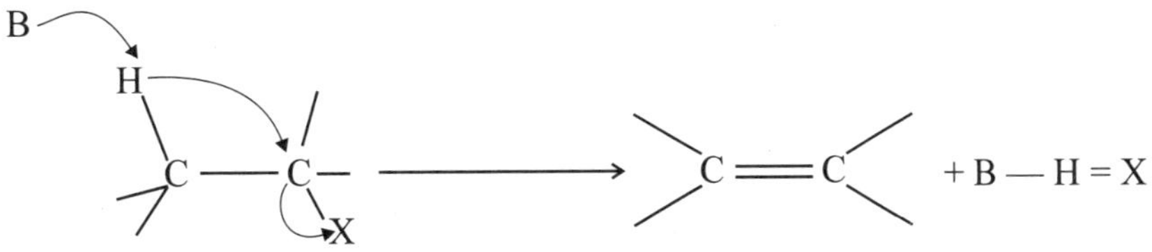

The mechanism is labeled bimolecular because both the base, B⁻, and alkyl halide are in the highest- energy transition state, the rate-determining step. This transition state in E2 dehydrohalogenation is as follows.

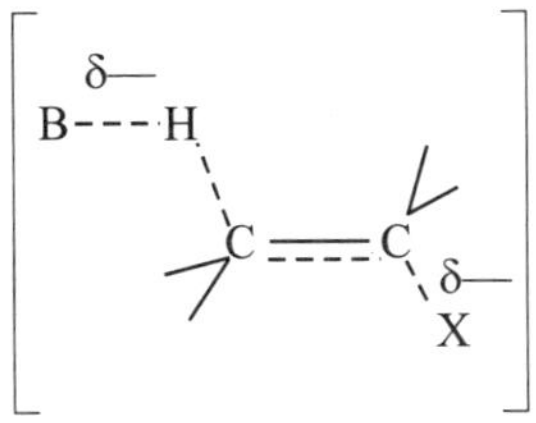

E2 Transition State

The E1 mechanism is a carbocation mechanism and is important in tertiary and some secondary alkyl halides. E1 stands for elimination unimolecular, which means that only one species is found in the transition state. The E1 mechanism goes through a carbocation. The first step of this mechanism, the rate-determining step, occurs when the halide and the pair of electrons that bond it to the C atom leave, resulting in the formation of the carbocation. In the second step, the base removes a proton and the alkene results. The following shows the E1 elimination of *t*-butyl bromide to produce methylpropene.

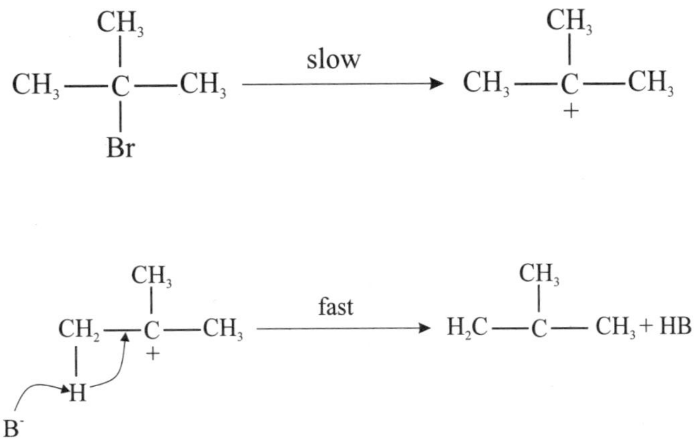

Test Yourself

1. a   Draw the structure of 3,4-dimethyl-1-pentene.

   b.   Does this alkene exist as *cis* and *trans* isomers? Explain.

2.   What is the IUPAC name for the following alkene?

3. Draw the structures of *cis-* and *trans*-2-methyl-3-hexene.

4. What is the name of the following molecule?

5. What is the name for the following?

6. What is incorrect regarding the name 2-butylcyclohexene?

7. a. What is the hybridization of the C atoms in a C–C double bond?
    b. What is the geometry of a C–C double bond?
    c. How does the π bond of the double bond form?
    d. Where is the π bond located?

8. Explain why *cis*-2-butene cannot change to *trans*-2-butene at 298 K and 1 atm.

9. Is the following molecule an *E* or *Z* isomer?

10. Draw the structure of (*Z*)-3-bromo-1,2-dichloro-2-pentene.

11. Rank the four alkene isomers of C_4H_8 from least stable to most stable. Explain your ranking.

12. a. Draw the structures of (*E*)-cyclooctene and (*Z*)-cyclooctene.

    b. Which of these structures has a higher enthalpy of hydrogenation? Explain.

13. a. What is the principal mechanism of an acid-catalyzed dehydration of an alcohol?

    b. What is the principal mechanism of the dehydrohalogenation of alkyl halides?

14. Draw the structures for the major products of the acid-catalyzed dehydration of 2-methylcyclohexanol. Label the major and minor products.

15. Write the IUPAC names of the two products of the acid-catalyzed dehydration of 3-pentanol and state which one is the major product.

16. When 2,2-dimethylcyclohexanol undergoes acid-catalyzed dehydration, it rearranges and forms a more stable alkene product.

    a. What is the rearranged product of this reaction?

    b. What product forms if it does not undergo rearrangement?

17. What are the structures of the products of the reaction of 2-bromo-2-methylhexane and sodium ethoxide? Label the major and minor products.

18. Draw the structure of the transition state for the E2 elimination reaction of 2-bromopropane in the presence of ethoxide, $CH_3CH_2O^-$.

19. Starting with cyclohexane, show a synthetic pathway that produces cyclohexene.

20. Which of the following alkenes is most stable?

    a. 1-heptene

    b. 1-methylcyclohexene

    c. 2-heptene

    d. 1,2-dimethylcyclohexene

21. Which of the following most readily undergoes E2 elimination with a strong base?

    a. 1-bromo-2,2-dimethylpropane

    b. 2-bromo-2-methylbutane

    c. 2-bromopentane

    d. 2-bromo-3-methylbutane

22. Consider the dehydration of 2-methyl-1-propanol. Which of the following carbocations results when a hydride shift occurs in this mechanism?

    a. $(CH_3)_2CHCH_2^+$

    b. $(CH_3)_2C^+CH_3$

    c. $CH_3CH_2CH^+CH_3$

    d. none of these

23. An alkyl bromide produces a single alkene when it reacts with sodium ethoxide and ethanol. This alkene undergoes hydrogenation and produces 2-methylbutane. What is the identity of the alkyl bromide?

    a. 1-bromo-2,2-dimethylpropane

    b. 1-bromobutane

    c. 1-bromo-2-methylbutane

    d. 2-bromo-2-methylbutane

24. Consider the following carbocations.

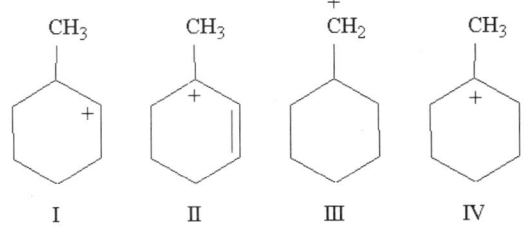

    I   II   III   IV

    Which of the following is the correct ranking from least stable to most stable?

    a. I < II < III < IV

    b. II < IV < I < III

    c. III < I < IV < II

    d. III < I < II < IV

25. What is the product(s) of the reaction of the following compound with alcoholic potassium hydroxide?

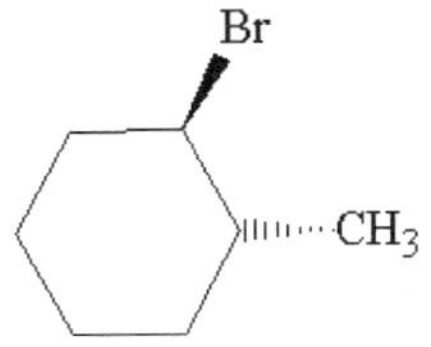

    a. 1-methylcyclohexene only

    b. 3-methylcyclohexene only

    c. 1-methylcyclohexene (major product), 3-methylcyclohexene (minor product)

    d. 3-methylcyclohexene (major product), 1-methylcyclohexene (minor product)

    e. none of these

26. Which of the following is an *E* isomer?

    a.

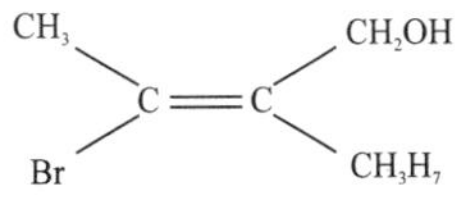

    b.

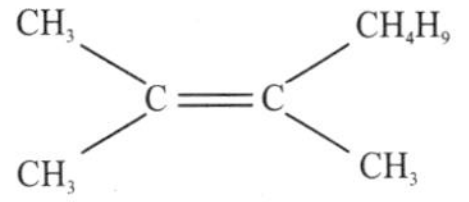

    c.

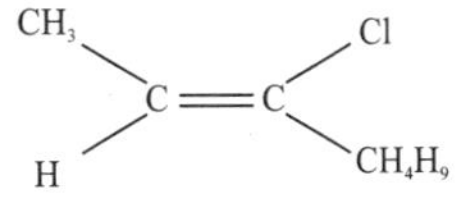

    d.

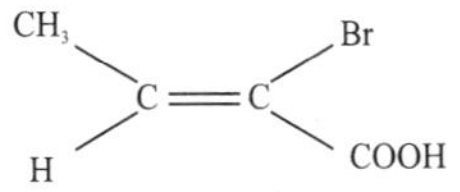

27. Which of the following has the highest
    enthalpy of hydrogenation?

    a. *(Z)*-4-methyl-2-pentene

    b. 2,4-dimethyl-2-hexene

    c. 1,2-diethylcyclopentene

    d. *(Z)*-2,2,5,5-tetramethyl-3-hexene

28. Which of the following undergoes E2
    elimination in the presence of a strong base to
    yield one product?

    a. 1-bromo-3,3-dimethylbutane

    b. 3-bromo-3-methylpentane

    c. 3-bromo-2-methylpentane

    d. 1-bromo-1-methylcyclohexane

29. Which of the following carbocations does not
    rearrange to a more stable carbocation?

    a.

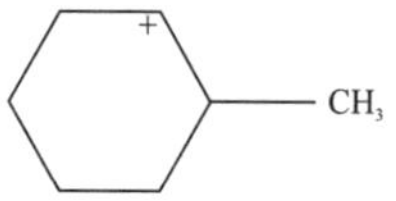

b.

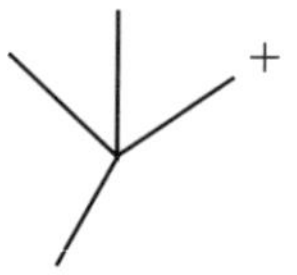

c.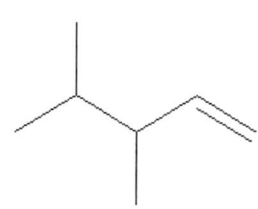

d. all of these will rearrange

30. Which of the following does not exist as
    geometric isomers?

    a. 3-bromo-1-chloro-1-pentene

    b. cyclodecene

    c. 3-bromo-2-methyl-2-butene

    d. 3-methyl-2-pentene

Check Yourself

1. a.

b. No, this does not exist as *cis* and *trans* isomers because the double bond is between the first and second C atoms. Two H atoms bond to the first C atom, and switching their positions produces the same structure. (**Alkene structures and isomers**)

2. Its name is 3,3-dichlorocyclohexene. In all cycloalkenes the double bond is always between the first and second C atoms. (**Alkene nomenclature**)

3.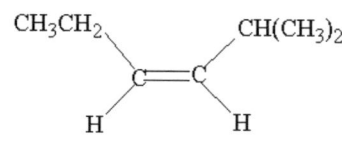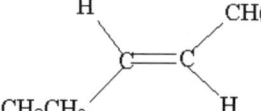

   cis-2-methyl-3-hexene   trans-2-methyl-3-hexene   (**Alkene structures and isomers**)

4. Its name is 1,2-dimethyl-3-vinylcyclopentane. A vinyl group is $H_2C=CH$. (**Alkene nomenclature**)

5. Its name is *trans*-8-methyl-4-nonene. (**Alkene nomenclature**)

6. Because the double bond is always between the first and second C atom in a cycloalkene, the numbering of the ring begins with the substituent group; therefore, the correct name for this compound is 1-butylcyclohexene. (**Alkene nomenclature**)

7. a. The C atoms in a double bond are sp^2 hybridized.
   b. The geometry at the double bond is trigonal planar, and the bond angles are 120°.
   c. The π bond results from the sideways overlap of the unhybridized p orbitals located on each C atom of the double bond.
   d. The π bond is located above and below the plane of the C atoms of the double bond. (**Alkene structure**)

8. The π bond in *cis*-2-butene must be broken for the rotation to occur that changes the molecule to *trans*-2-butene. At room conditions, 298 K and 1 atm, this system does not have enough energy for this to take place. (**Alkene structure**)

9. Because the higher priority groups are on opposite sides, it is the *E*-isomer of this molecule.

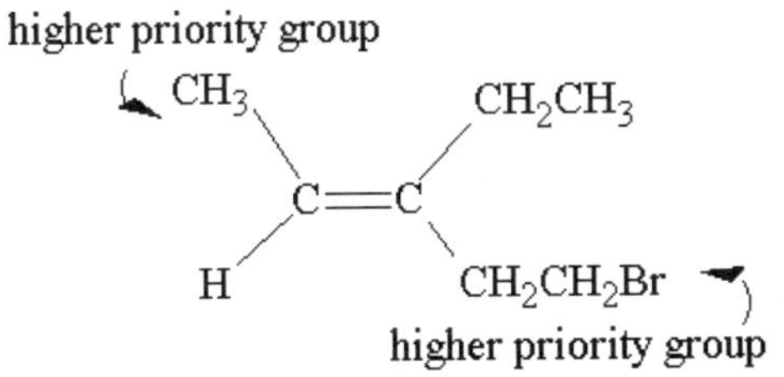

higher priority group

(***E-Z nomenclature***)

10.

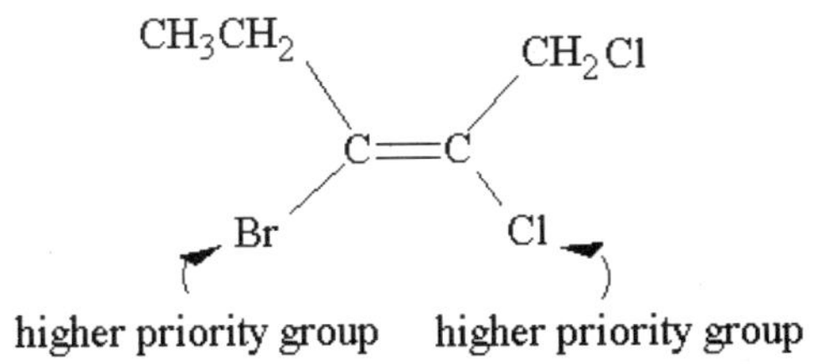

higher priority group   higher priority group

The *Z*-isomer has both higher priority groups on the same side of the molecule. (***E-Z nomenclature***)

11. 1-butene<*cis*-2-butene<*trans*-2-butene<2-methylpropene. The more highly substituted alkenes are more stable than those that are less highly substituted. *Trans*-isomers are more stable than *cis*-isomers. (**Alkene stability**)

12. a.

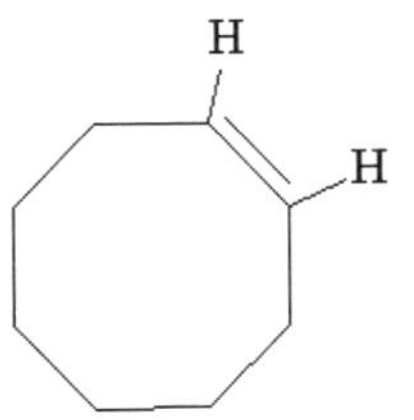

(Z)-cyclooctene 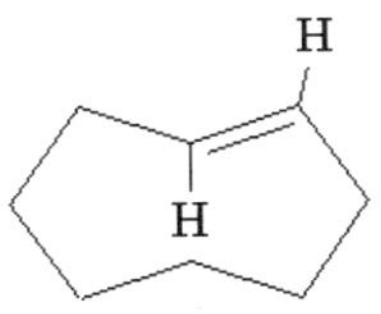   (E)-cyclooctene

b. The enthalpy of hydrogenation is significantly higher for the (*E*)-cyclooctene because the number of C atoms in the ring is too small to accommodate the *trans*-double bond. (**Alkene stability**)

13. a. Dehydration of alcohols is most likely an E1 reaction.
    b. Dehydrohalogenation of alkyl halides principally follows the E2 mechanism. **(Dehydrohalogenation)**

14.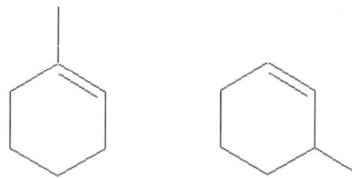

      1-methylcyclohexene   3-methylcyclohexene
        major product   minor product

The major product of this reaction is the more highly substituted one; i.e., it follows Saytzeff's rule. **(Dehydration of alcohols)**

15. The two products of the reaction are *cis*-2-pentene and *trans*-2-pentene. Because the *trans*-product is more stable than the *cis*-product, *trans*-2-pentene is the major product of this reaction. **(Dehydration of alcohols)**

16. **(Dehydration of alcohols)**

  a. 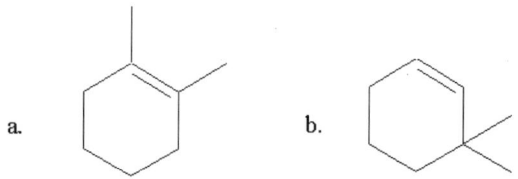   b.

    1,2-dimethylcyclohexene   3,3-dimethylcyclohexene

17. **(Dehydrohalogenation)**

        2-methyl-2-hexene   2-methyl-1-hexene
         major product   minor product

18. E2 elimination is concerted, with O–H bond formation, double bond formation, and C–Br bond breaking occurring at the same time. **(Dehydrohalogenation)**

$$\left[\begin{array}{c} CH_3CH_2O \overset{\delta-}{\cdots\cdots}H \\ H\diagup \overset{C = C}{\underset{H}{|}} \diagdown \overset{CH_3}{\underset{Br}{\delta-}} \end{array} \right]$$

19.

cyclohexane · chlorocyclohexane · cyclohexene

Cl_2 / light → NaOCH$_2$CH$_3$ / CH$_3$CH$_2$OH →

(Alkene reactions)

20. d. 1,2-Dimethylcyclohexene is the most stable alkene because it is the only tetrasubstituted alkene. **(Alkene stability)**

21. b. 2-Bromo-2-methylbutane is the only listed tertiary alkyl bromide. Tertiary alkyl bromides are usually most reactive in E2 mechanisms. **(Dehydrohalogenation)**

22. b. $(CH_3)_2C^+CH_3$, a tertiary carbocation results when a hydride shift occurs in this reaction. **(Dehydration of alcohols)**

23. c. 1-Bromo-2-methylbutane undergoes dehydrohalogenation and produces 2-methyl-1-butene, which undergoes hydrogenation to produce 2-methylbutane.**(Alkene reactions)**

24. c. III < I < IV < II is the correct order from least to most stable. The least stable carbocation is primary, I. Next in stability is the secondary carbocation, I. The remaining two cations are tertiary, but II is more stable because it is an allylic cation. **(Carbocations)**

25. b. 3-Methylcyclohexene is the only product of the reaction because of the position of the methyl group. For E2 dehydrohalogenation to occur, the H atom that is removed must be anti-coplanar to the halogen atom. The methyl group prevents the removal of the H atom that would produce 1-methylcyclohexene. **(Dehydrohalogenation)**

26. a. The *E*-isomer is the *trans*-isomer, which means that the high-priority groups are on opposite sides of the double bond. The Br atom is the higher-priority group on one side, and the CH$_2$OH group is the higher-priority group on the other side. **(*E-Z* isomers)**

27. d. *(Z)*-2,2,5,5-Tetramethyl-3-hexene has the highest enthalpy of hydrogenation because it is disubstituted with two *t*-butyl groups on the same side of the molecule. These groups repel each other, increasing the energy of the molecule. **(Alkene stability)**

28. a. 1-Bromo-3,3-dimethylbutane only has one adjacent C atom with a H atom that can be abstracted; thus, it only produces one product, 3,3-dimethyl-butene. **(Dehydrohalogenation)**

29. d. All of these carbocations can rearrange to a more stable one. **(Carbocations)**

30. c. 3-Bromo-2-methyl-2-butene cannot exist as *cis-trans* isomers. **(Alkene isomers)**

Grade Yourself

Circle the number of questions you missed, then fill in the total incorrect for each topic. If you answered more than three questions incorrectly, you need to focus on that topic. (If a topic has less than three questions and you had at least one wrong, we suggest you study that topic also. Read your textbook, a review book, or ask your teacher for help.)

Subject: Alkenes—Structure, Stability, and Preparations

Topic	Question Numbers	Number Incorrect
Alkene structures and isomers	1, 3, 7, 8, 30	
Alkene nomenclature	2, 4, 5, 6	
E-Z nomenclature	9, 10	
Alkene stability	11, 12, 20, 27	
Dehydrohalogenation	13, 17, 18, 21, 25, 28	
Dehydration of alcohols	14, 15, 16, 22	
Alkene reactions	19, 23	
Carbocations	24, 29	
E-Z isomers	26	

Alkenes Reactions— Electrophilic Addition Reactions

5

Brief Yourself

Addition reactions are the principal reactions that alkenes undergo. One type of addition reaction occurs when a symmetrical reagent such as H_2, Cl_2, or Br_2 breaks the π bond of the double bond and attaches at that position. An example of such a reaction is the hydrogenation of alkenes.

$$\text{alkene} \quad \overset{\text{Hydrogenation reaction}}{\longrightarrow} \quad \text{alkane}$$

The second type of addition reaction occurs when an unsymmetrical reagent adds across the double bond. If the attacking molecule has the general formula of E–X, this reaction occurs as follows.

In this reaction, E is the electrophilic or the more electropositive atom and X is the more electronegative atom. Examples include hydrohalogenation, adding HX; hydration, adding H_2O (HOH); and the addition of sulfuric acid, H_2SO_4.

Hydrohalogenation occurs when a hydrogen halide such as HCl, HBr, or HI adds to an alkene and produces an alkyl halide. An example of a hydrohalogenation reaction is the combination of hydrogen chloride and cyclopentene, producing chlorocyclopentane.

cyclopentene

chlorocyclopentane

The H from HCl bonds to one C atom of the double bond and Cl bonds to the other. This follows a carbocation mechanism in which the electrophile, H+, first attacks the π bond of the double bond (an electron-rich region), producing a carbocation. This is the rate-determining step. The second step is the attack of the chloride ion, producing the alkyl halide. The name of this mechanism is electrophilic addition. Rearrangements may occur because this reaction produces a carbocation intermediate.

In the electrophilic addition of hydrogen halides to alkenes in which the C atoms of the double bond are not bonded to the same groups, only one of two expected products is observed, another example of a regiospecific reaction. In the hydrohalogenation of alkenes, the H atom bonds to the C atom with the most H atoms. This is a statement of Markovnikov's rule. A modern statement of Markovnikov's is that the product that results from the electrophilic addition of an unsymmetrical reagent to an alkene is the one that results from the most stable carbocation.

If pure HBr reacts with 1-methylcyclohexene, the only product of the reaction is the expected Markovnikov product 1-bromo-1-methylcyclohexane. If the same reaction is run with the addition of an organic peroxide, ROOR, the anti-Markovnikov product is produced. The addition of the organic peroxide causes the reaction to go through a free-radical mechanism.

The hydrogen halides, HX, are acidic substances; therefore, other acids can undergo electrophilic addition to double bonds. Sulfuric acid undergoes this reaction and follows a similar mechanism through a carbocation. For example, consider the reaction of sulfuric acid and methylpropene.

methylpropene $+$ H_2SO_4 $\longrightarrow$

OSO_2OH

t-butyl hydrogen sulfate

The Markovnikov product of *t*-butyl hydrogen sulfate results. Heating this product with water hydrolyzes its O–S bond and *t*-butyl alcohol forms.

In a hydration reaction, alkenes react with water in the presence of an acid to produce alcohols. This is another example of an electrophilic addition reaction. For example, consider the hydration reaction of 1-ethylcyclobutene, producing 1-ethylcyclobutanol. The mechanism for this reversible reaction is the opposite to that of the dehydration of 1-ethylcyclobutanol.

Another way to produce alcohols from alkenes is through oxymercuration-demercuration. This is a two-step reaction that begins with an alkene, mercury(II) acetate, and water, and produces hydroxyalkyl mercury(II) acetate and acetic acid. This is the oxymercuration step. It is followed by demercuration in which the hydroxyalkyl mercury(II) acetate is treated with sodium borohydride, $NaBH_4$, (a strong reducing agent) and base. For example, consider the oxymercuration-demercuration of 2-methyl-1-pentene, which produces 2-methyl-2-pentanol.

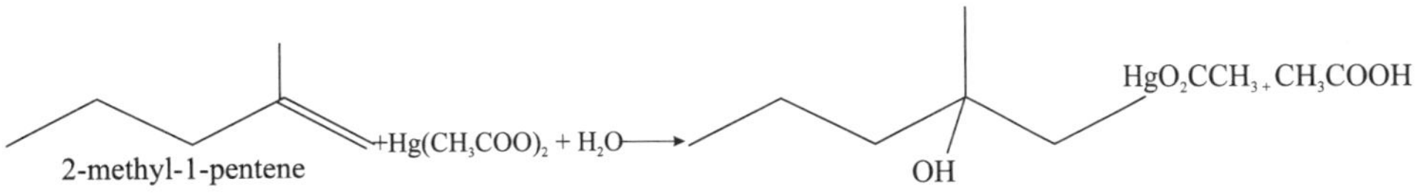

Oxymercuration

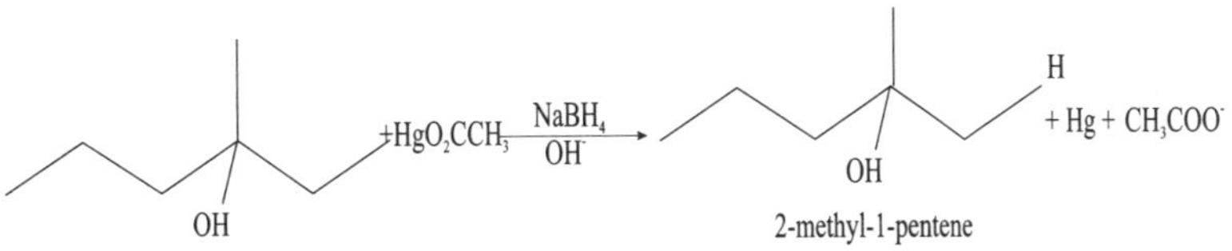

Demercuration

Tetrahydrofuran, THF, is often used as the solvent for this reaction. The product of oxymercuration-demercuration reactions is the Markovnikov product because it is an electrophilic addition.

A third method to produce alcohols from alkenes is hydroboration-oxidation. The alkene is first treated with diborane, B_2H_6, in a solvent such as diglyme (diethylene glycol dimethyl ether). This reaction produces a trialkyl borane, R_3B. For example, consider the reaction of ethene with diborane.

ethene   ethene   triethylborane

In the second step, the trialkyl borane is oxidized with hydrogen peroxide, H_2O_2, in base, OH^-. Ethanol, CH_3CH_2OH, is the product of this reaction. Note that hydroboration-oxidation has the opposite regioselectivity to that of acid-catalyzed hydration and oxymercuration-demercuration. Hydroboration-oxidation produces the anti-Markovnikov product. It is also stereospecific. Both the H and OH groups added to the molecule through hydroboration-oxidation are on the same side of the resulting alcohol. For example, consider the hydroboration-oxidation of 1-methylcyclohexene.

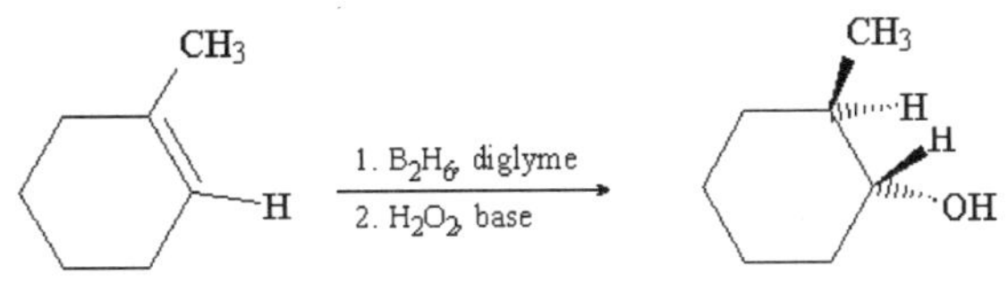

1-methylcyclohexene   *trans*-2-methycyclohexanol

The product, *trans*-2-methylcyclohexanol, has the OH group on the C atom adjacent to the methyl group; thus, it is the anti-Markovnikov product. The H atom and OH group added during the reactions are on the same side of the molecule. This is a syn addition.

When halogens react with alkenes, they produce 1,2-dihaloalkanes. For example, when propene reacts with bromine, Br_2, in carbon tetrachloride, the product is 1,2-dibromopropane.

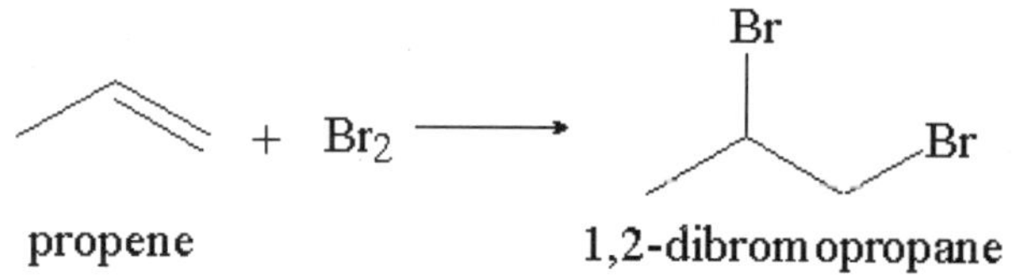

propene   + Br_2 ⟶   1,2-dibromopropane

The mechanism for this reaction goes through a cyclic bromonium ion intermediate. First the π bond of the alkene attacks the Br–Br bond as follows.

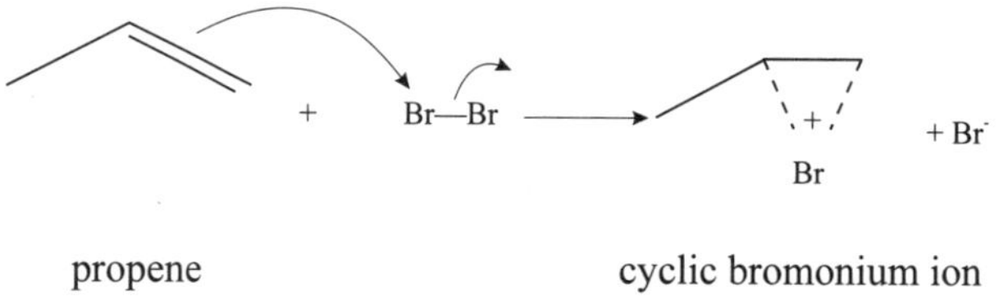

propene   cyclic bromonium ion

Then the Br^- attacks the bromonium ion as follows, and yields the product.

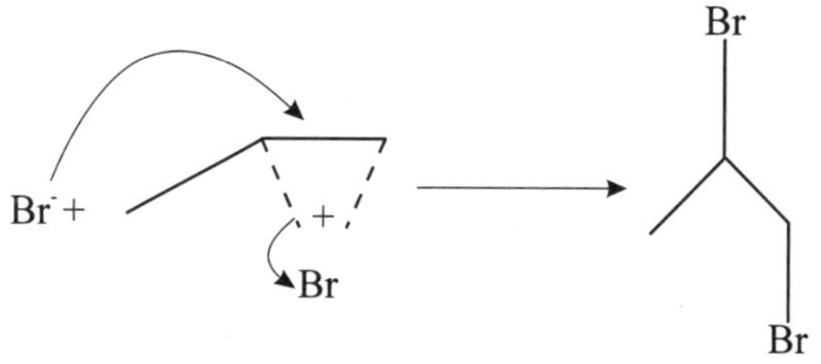

This is an anti addition, which means that the Br atoms attach from different sides of the molecule. If cyclopentene and Br_2 in carbon tetrachloride react, the product is *trans*-1,2-dibromocyclopentane.

If halogenation is carried out in water instead of a nonpolar solvent such as carbon tetrachloride, a halohydrin results. A halohydrin is a compound that has an OH group on a C atom adjacent to a C atom bonded to a halogen atom. The mechanism for halohydrin formation is similar to that of halogenation. Initially, the halogen attacks the π system, producing the cyclic halonium ion. This is followed by the attack of the water molecule that opens the ring. When the ring opens, the C atom that can best accommodate the positive charge is where the water bonds.

An epoxide, sometimes called an oxacyclopropane, is a compound that has a three-membered ring with two C atoms and one O atom. The simplest epoxide is ethylene oxide (oxirane–IUPAC name).

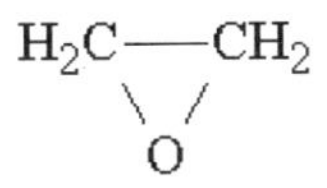

$$H_2C \overset{\diagdown \diagup}{\underset{O}{\text{——}}} CH_2$$

ethylene oxide

oxirane

A common way to name epoxides is to derive their names from the alkane chain in which the ring is located. For example, consider the structure of 2-ethyl-2,3-epoxyhexane.

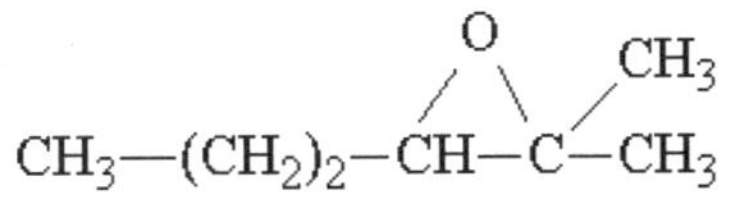

$$CH_3-(CH_2)_2-CH-C-CH_3$$

2-methyl-2,3-epoxyhexane

This structure has a chain of six C atoms with the epoxide ring at the second and third C atoms. It also has a methyl group on the second C atom.

Alkenes react with peracids, RCO_3H, and produce epoxides.

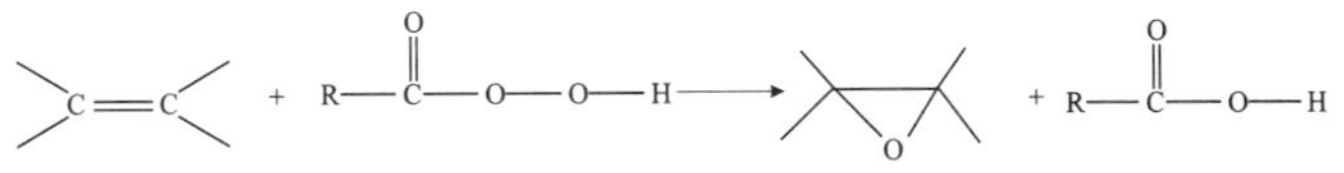

alkene   peracid   epoxide   carboxylic acid

Epoxides are stable compounds unless they are in an aqueous acidic or basic solution. Under these conditions the ring opens and forms a glycol, a 1,2-diol.

When alkenes react with ozone, O_3, followed by a reducing agent such as dimethylsulfide, they produce carbonyl compounds—aldehydes and ketones. Ozone initially attacks an alkene and produces an unstable ozonide, which undergoes reduction with $CH_3–S–CH_3$ and produces two carbonyl compounds.

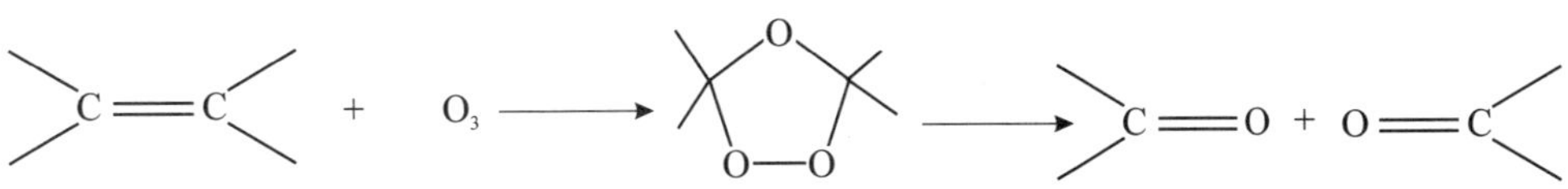

alkene   ozone   ozonide   two carbonyl compounds

These two reactions, ozonide formation followed by reduction, are called ozonolysis. For example, consider the ozonolysis of 2-methyl-1-butene.

| 2-methyl-1-butene | 1. O₃ / 2. CH₃SCH₃ → | 2-butanone | + | HCHO formaldehyde |

Alkene molecules can also be cleaved by treatment with concentrated potassium permanganate, KMnO₄, followed by treatment with acid. This reaction yields ketones, carboxylic acids, and/or carbon dioxide. For example, consider the reaction of 3-methyl-2-pentene with KMnO₄ and then H⁺.

3-methyl-2-pentene   1. KMnO₄ 2. H⁺ →   2-butanone   + CH₃—C—OH ethanoic acid

Alkenes are reactive enough that they can react with themselves. Under the proper conditions, two alkene molecules react and produce a dimer, a molecule that results from the combination of two like molecules. If A is an alkene molecule, then A–A is a dimer. The starting alkene molecule is termed the monomer. Three alkene monomers, A, combine to produce a trimer, A–A–A. This reaction can continue, producing long C chains. These large molecules, macromolecules, are called polymers. For example, *n* ethene molecules, H₂C=CH₂, react under the proper conditions to produce polyethylene.

n H₂C=CH₂ —polymerization→ —(CH₂CH₂)ₙ—

ethene   polyethylene

Both carbocation and free-radical mechanisms are used in polymerization reactions.

Test Yourself

1. a. What is the general type of reaction that alkenes undergo?

   b. Why are these reactions given this name?

2. a. Write the equation for the reaction of hydrogen chloride and 2-methyl-1-butene.

   b. Write the name(s) for the product(s) of the reaction.

3. Consider the reaction of hydrogen bromide and 1-ethylcyclopentene. Draw the structure of the most and least stable carbocations that form during this reaction.

4. a. Write the name of the principal product that forms when hydrogen bromide reacts with 1-heptene.

   b. Write the name of the principal product that forms when hydrogen bromide reacts with 1-heptene in the presence of an organic peroxide.

   c. What type of mechanism occurs in the reaction in part a?

   d. What type of mechanism occurs in the reaction in part b?

5. a. What is the product for the following reaction?

   =CH$_2$ + H$_2$SO$_4$

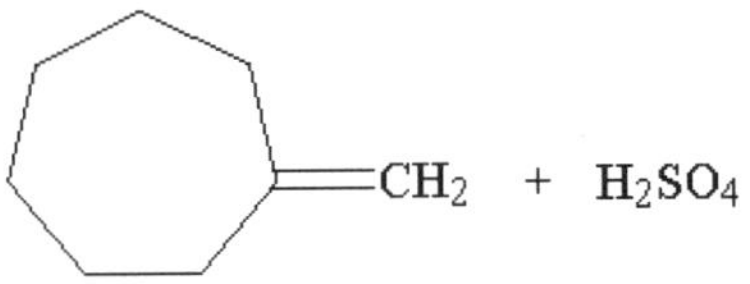

   b. What compound results when the product of this reaction is hydrolyzed?

6. Write the names of two alkenes that undergo hydration reactions to produce the following alcohol.

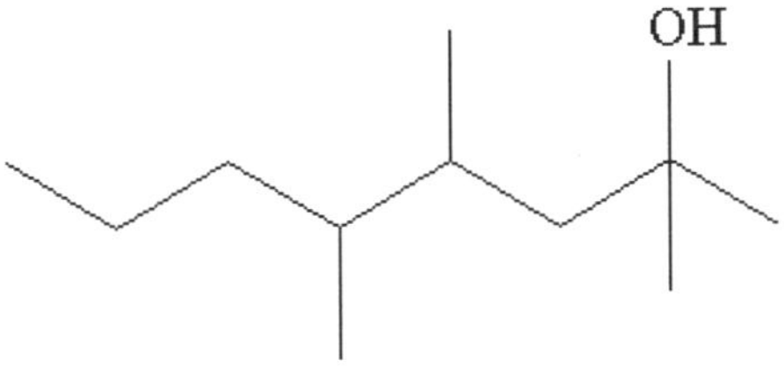

7. What is the product of the following reaction?

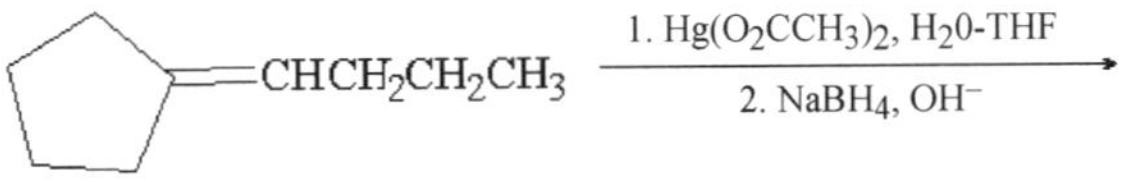

   =CHCH$_2$CH$_2$CH$_3$ $\xrightarrow[\text{2. NaBH}_4\text{, OH}^-]{\text{1. Hg(O}_2\text{CCH}_3)_2\text{, H}_2\text{0-THF}}$

8. Draw the structure of the lowest-energy carbocation produced when 2-methyl-1-pentene undergoes oxymercuration-demercuration.

9. Draw the structure of the product of the following reaction.

   $\xrightarrow[\text{2. H}_2\text{O}_2\text{, OH}^-]{\text{1. B}_2\text{H}_6\text{, diglyme}}$

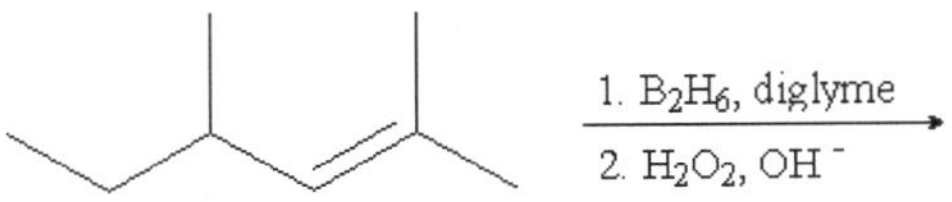

10. Draw the structure of the product that results when the following alkene undergoes hydroboration-oxidation. Hint: Do not forget stereochemical considerations.

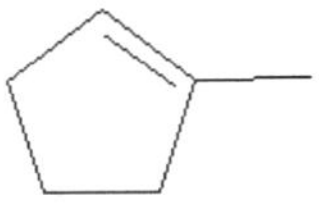

11. Draw the structure and name the product of the reaction of cycloheptene and bromine in carbon tetrachloride. Do not forget stereochemical considerations.

12. Write an equation for a reaction that produces 1-chloro-2-methyl-2-hexanol.

13. Show how 2,3-dimethyl-2,3-epoxypentane may be synthesized.

14. Draw the structure(s) of the ozonolysis product(s) of 3-methylcyclohexene.

15. Write the equation for the reaction of 3-methyl-2-heptene with potassium permanganate followed by acid hydrolysis.

16. a. Write an equation that shows the polymerization reaction of vinyl chloride.

   b. What is the name of the product of this reaction?

17. Outline a synthetic pathway that produces 2-bromo-1-methylcyclopentane starting with 1-methylcyclopentanol and any necessary reagents.

18. Draw the structure of the most significant reaction intermediate in the mechanism for the bromination of cyclohexene without peroxides.

19. When 1,6-dimethylcyclohexene reacts with HCl, two products result that have the formula of C$_8$H$_{15}$Cl. What are the structures of these products? Explain the formation of these products in terms of the reaction mechanism.

20. Which of the following reactions does not undergo electrophilic addition to a double bond?

   a. hydrohalogenation

   b. hydroboration-oxidation

   c. epoxidation

   d. all of the above undergo electrophilic addition

21. Which of the following reagents converts 1-butene to 2-butanol in greatest yield?

    a. 1. BH_3, THF-H_2O, 2. H_2O_2, OH

    b. 1. $Hg(O_2CCH_3)_2$, H_2O, 2. $NaBH_4$, OH^-

    c. 1. O_3, 2. Zn, H_2O

    d. H^+, H_2O

    e. none of these

    (For questions 22 and 23 consider the structure of limonene, a compound that gives the characteristic odor and flavor to lemons.)

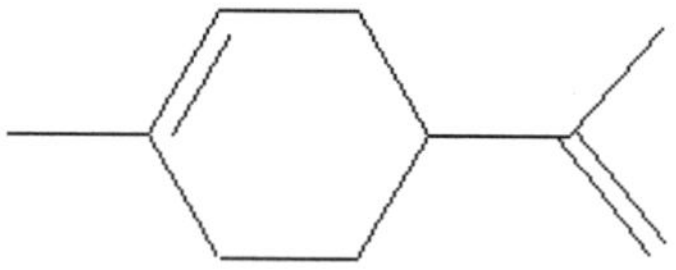

    limonene

22. What product results from the reaction of limonene and chlorine water?

    a.

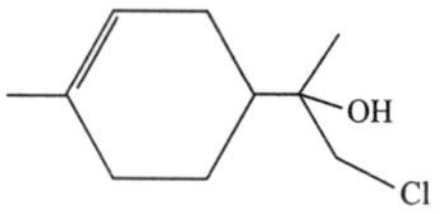

    b.

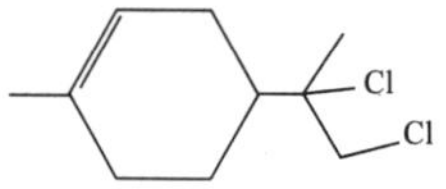

    c.

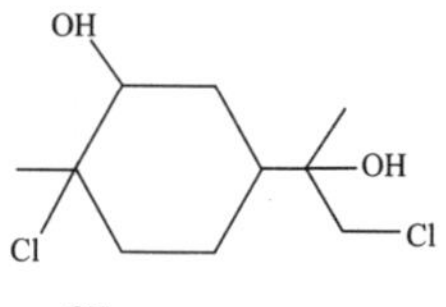

    d.

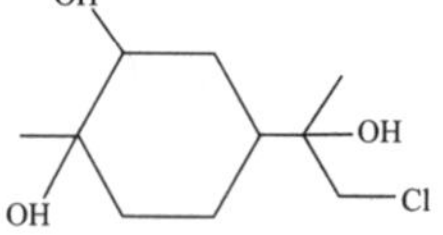

    e. none of these

23. What product results from the complete hydroboration-oxidation of limonene?

    a.

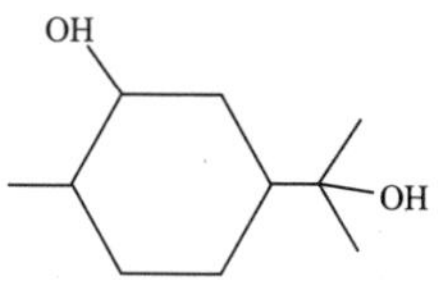

    b.

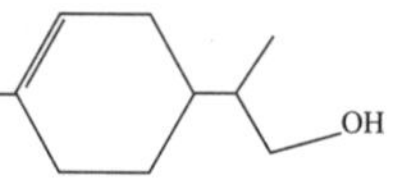

    c.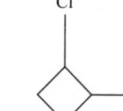

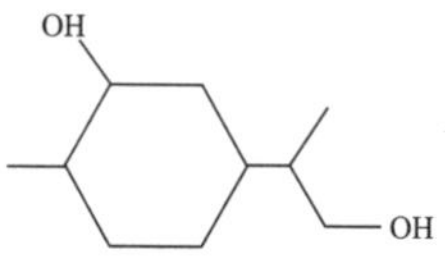

    d.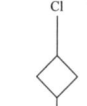

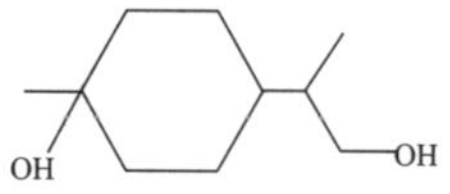

    e. none of these

24. An unknown compound W has the formula C_5H_9Cl. It does not react with bromine in carbon tetrachloride. When treated with a strong base it produces a single compound, X, with the formula C_5H_8 that reacts with bromine in carbon tetrachloride. Ozonolysis of compound X, using ozone followed by dimethylsulfide, produces a compound with the formula $C_5H_8O_2$. Which of the following is the structure of W?

    a.

    b.

    c.

    d.

    e. none of these

25. Consider the following alkene.

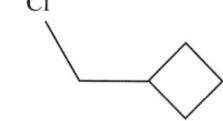

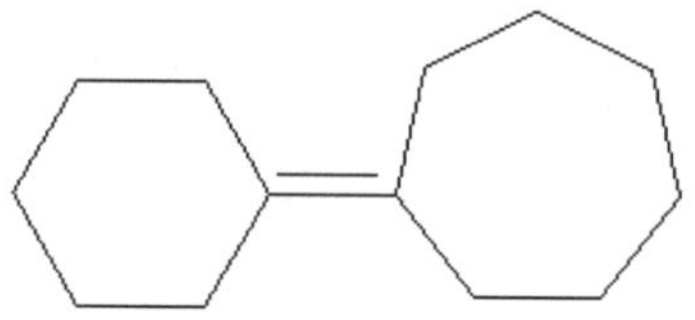

Considering only the following compounds, select the product(s) of the ozonolysis (treatment with ozone followed by dimethylsulfide) of this compound.

I.

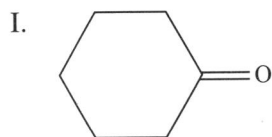

II.

III.

O

H

IV.

OH

a. I and II

b. III and IV

c. I and IV

d. II and III

e. none of these

26. Which of the following has the highest reaction rate when treated with bromine in a nonpolar solvent?

a. ethene

b. propene

c. 2-methylpropene

d. 2,3-dimethyl-2-butene

27. Which of the following is **not** a correct statement concerning the mechanism for hydroboration-oxidation?

a. Borane forms a π complex with the alkene.

b. A three-centered two electron bond joins two C atoms and one B atom as borane bonds to the alkene.

c. Three alkene molecules react with one borane.

d. All of the above are correct statements.

28. What is the product of propene with cold sulfuric acid?

a. $CH_3CH_2CH_2OH$

b. $CH_3CH_2CH_2OSO_3H$

c. $CH_3CH(OSO_3H)CH_3$

d. $CH_3CH(OH)CH_3$

e. none of these

29. Which of the following is the principal product of the reaction of 3-methyl-1-butene with HCl?

a. 2-chloro-3-methylbutane

b. 2-chloro-2-methylbutane

c. 1-chloro-3-methylbutane

d. none of these

30. Which of the following is **not** a product that can be obtained from the oxidation of alkenes with potassium permanganate followed by acid hydrolysis?

a. ketones

b. aldehydes

c. carboxylic acids

d. CO_2

✔ Check Yourself

1. a. Many alkene reactions are classified as electrophilic addition reactions.

b. An electrophilic addition reaction occurs when an electron-deficient species, an electrophile (E) attacks and breaks the π bond of the C–C double bond. Then, the nucleophilic part of the attacking molecule (X) bonds to the position adjacent to that of the electrophile. (**Electrophilic addition**)

2. **(Hydrohalogenation)**

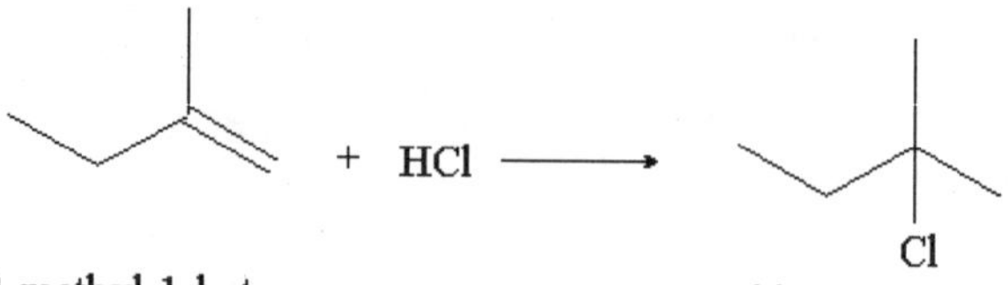

2-methyl-1-butene   + HCl ⟶   2-chloro-2-methylbutane

3. **(Hydrohalogenation)**

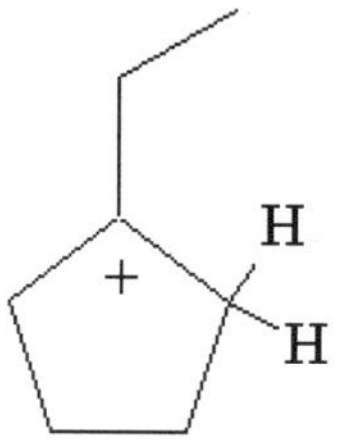

most stable
(tertiary carbocation)

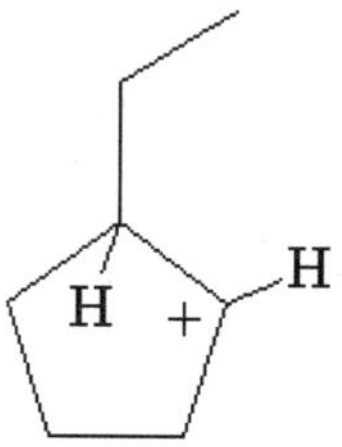

least stable
(secondary carbocation)

4. a. The product is 2-bromoheptane.

b. The product is 1-bromoheptane.

c. The reaction follows electrophilic addition through a carbocation intermediate.

d. The reaction follows a free-radical mechanism.

5. **(Hydration reactions)**

a.

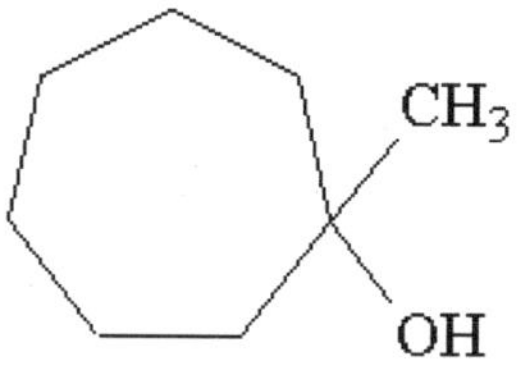

b.

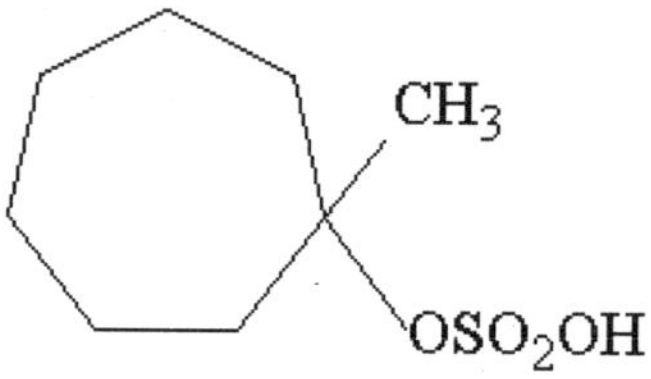

CH_3

OH

CH_3

OSO_2OH

6. Either 2,4,5-trimethyl-1-octene or 2,4,5-trimethyl-2-octene undergoes hydration reactions to produce 2,4,5-trimethyl-2-octanol. **(Hydration reactions)**

7. The oxymercuration-demercuration reaction is a hydration reaction that gives the Markovnikov product. **(Oxymercuration-demercuration)**

$—CH_2CH_2CH_2CH_3$

OH

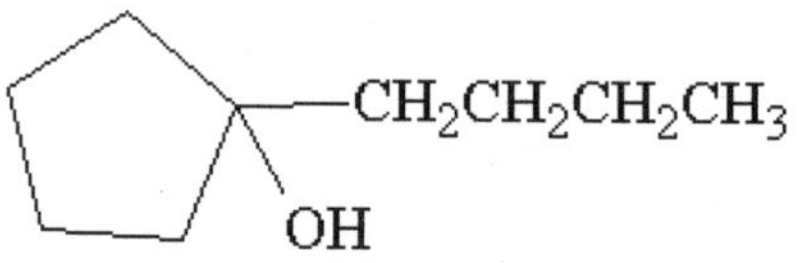

8. The carbocation that results is a cyclic mercury-substituted cation, as shown below.

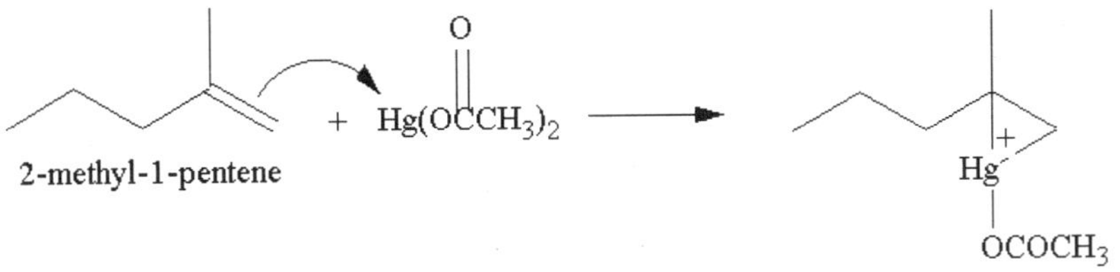

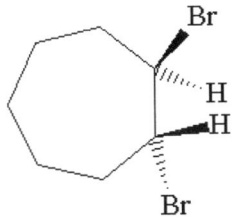

2-methyl-1-pentene   +   Hg(OCCH₃)₂ ⟶

OCOCH₃

(Oxymercuration-demercuration)

9. Hydroboration-oxidation gives exclusively the anti-Markovnikov hydration product.

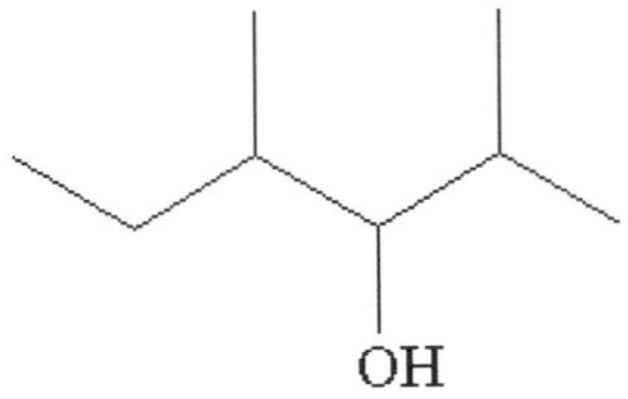

OH

(Hydroboration-oxidation)

10. The product is *trans*-2-methylcyclopentanol because the H atom and OH group bond to the same side of the molecule, a syn addition.

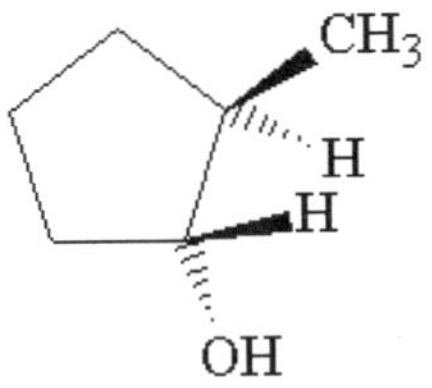

CH₃

H

H

OH

(Hydroboration-oxidation)

11.

Br

H

H

Br

trans-1,2-dibromocycloheptane

(Halogenation)

12.

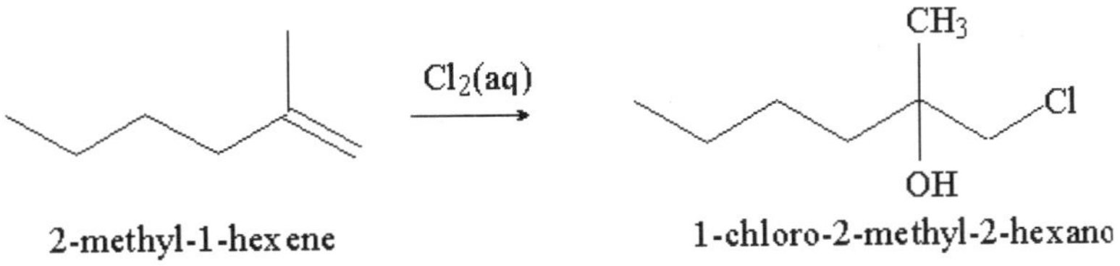

2-methyl-1-hexene $\xrightarrow{Cl_2(aq)}$ CH₃ ... Cl

OH

1-chloro-2-methyl-2-hexano

(Halohydrin formation)

13.

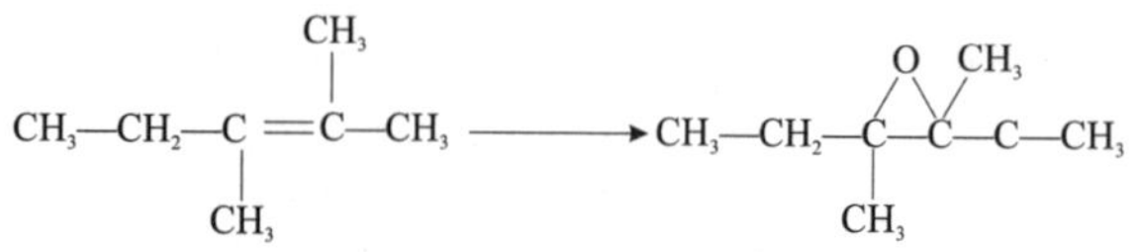

CH₃—CH₂—C(CH₃)=C(CH₃)—CH₃ ⟶ CH₃—CH₂—C(CH₃)—C(O)(CH₃)—C—CH₃

$$CH_3-CH_2-\underset{\underset{CH_3}{|}}{C}=\underset{|}{C}-CH_3 \longrightarrow CH_3-CH_2-\underset{\underset{CH_3}{|}}{C}-\overset{O}{\overset{\diagdown}{C}}-C-CH_3$$

2, 3-dimethyl-2pentene   2, 3-dimethyl-2, 3 epoxypentane

(Epoxidation)

14.

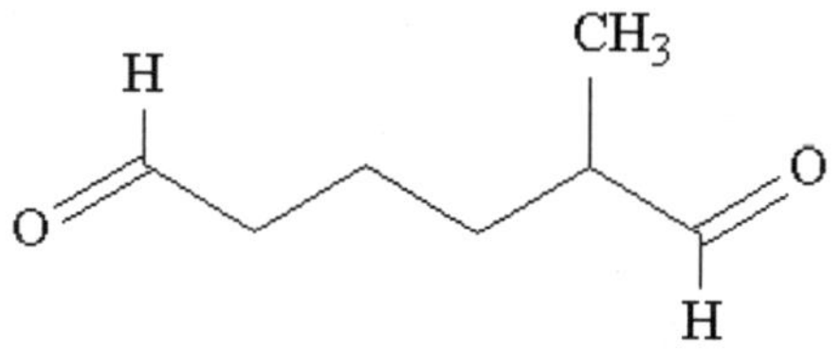

(Ozonolysis)

15.

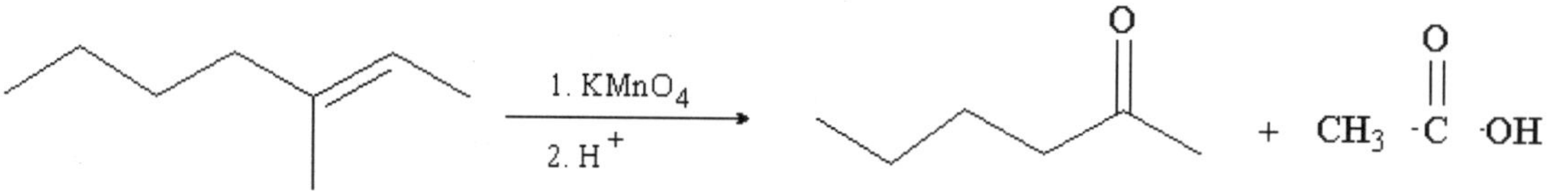

3-methyl-2-heptene $\xrightarrow[\text{2. H}^+]{\text{1. KMnO}_4}$ 2-heptanone $+$ $CH_3-\overset{O}{\overset{\|}{C}}-OH$ ethanoic acid

3-methyl-2-heptene   2-heptanone   ethanoic acid

(Alkene oxidation)

16. $n\ H_2C=CH-Cl \xrightarrow{\text{polymerization}} -(CH_2CH)_n-$ Cl

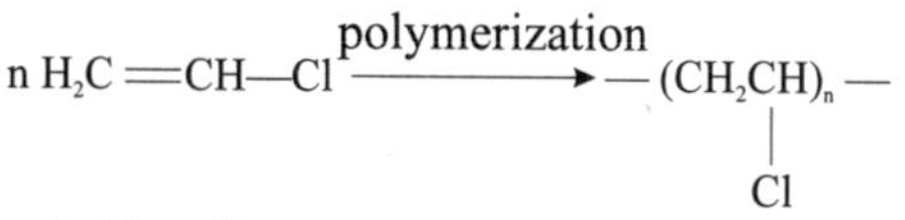

1-chloroethene
vinyl chloride   polyvinyl chloride

(Polymerization)

17. First dehydrate the alcohol to an alkene, and then add HBr with peroxide to attach the Br atom to the proper C atom.

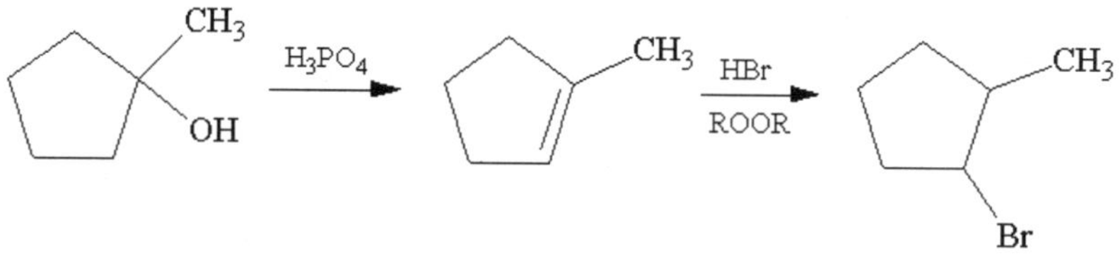

$\xrightarrow{\text{H}_3\text{PO}_4}$ $\xrightarrow[\text{ROOR}]{\text{HBr}}$

(Alkene reactions)

18.

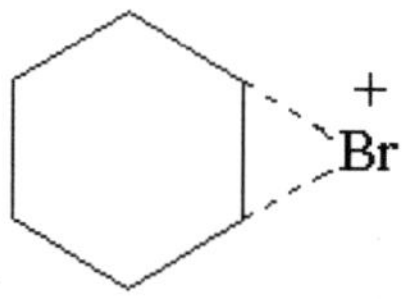

cyclic bromonium ion

In this mechanism, the Br_2 molecule attacks the π bond and produces a cyclic bromonium ion, which is then attacked by the remaining Br^- ion. (**Halogenation**)

19.

Hydrochlorination produces a carbocation intermediate that results from the attack of the H^+. The Cl^- attacks the trigonal planar carbocation from either the top or bottom. This means that the Cl and H atoms from HCl can be either on the same or opposite sides; thus, it produces two geometric isomers. (**Hydrohalogenation**)

20. d. Hydrohalogenation, hydroboration-oxidation, and epoxidation are electrophilic addition reactions. (**Alkene reactions**)

21. b. 1. $Hg(O_2CCH_3)_2$, H_2O, 2. $NaBH_4$, OH^-. The best yields for converting alkenes to alcohols are obtained by using oxymercuration-demercuration if the Markovnikov product is needed. (**Alkene reactions**)

22. d. This reaction forms a chlorohydrin at both double bonds. In the formation of the chlorohydrin, the OH group bonds to the tertiary position. (**Halohydrin formation**)

23. c. Hydroboration-oxidation places the OH group in the anti-Markovnikov position. (**Hydroboration-oxidation**)

24. b. 1-Chloro-3-methylcyclobutane is the only one of the listed compounds that fits the description. Another compound (not listed) that has these same chemical properties is chlorocyclopentane. (**Alkene reactions**)

25. a. I and II are the ketones that result from the ozonolysis of the compound. (**Ozonolysis**)

26. d. 2,3-Dimethyl-2-butene reacts significantly faster than the others because it has four R groups bonded to the double bond. These electron-releasing R groups help stabilize the cyclic bromonium ion that forms in the mechanism. (**Halogenation**)

27. d. All of the statements are correct regarding the mechanism of hydroboration-oxidation. (**Hydroboration-oxidation**)

28. c. $CH_3CH(OSO_3H)CH_3$ is the product of the reaction of propene and cold sulfuric acid. In this reaction the sulfuric acid adds across the double bond and produces an alkyl hydrogen sulfate. (**Hydrogen sulfate formation**)

29. b. 2-Chloro-2-methylbutane is the major product of this reaction because a 1,2-hydride shift occurs, producing a more stable tertiary carbocation. (**Hydrohalogenation**)

30. b. Aldehydes oxidize to acids in this reaction because $KMnO_4$ is a strong oxidizing agent. (**Alkene oxidations**)

Grade Yourself

Circle the number of questions you missed, then fill in the total incorrect for each topic. If you answered more than three questions incorrectly, you need to focus on that topic. (If a topic has less than three questions and you had at least one wrong, we suggest you study that topic also. Read your textbook, a review book, or ask your teacher for help.)

Subject: Alkenes Reaction—Electrophilic Addition Reactions

Topic	Question Numbers	Number Incorrect
Electrophilic addition	1	
Hydrohalogenation	2, 3, 19	
Hydration reactions	6	
Oxymercuration-demercuration	7, 8	
Hydroboration-oxidation	9, 10, 23, 27	
Halogenation	11, 18, 26	
Epoxidation	13	
Ozonolysis	14, 25	
Alkene oxidation	15, 30	
Polymerization	16	
Alkene reactions	17, 20, 21, 24	
Hydrogen sulfate formation	28	

Stereochemistry

 ## Brief Yourself

Stereochemistry is the study of the three-dimensional structure of molecules. Stereoisomers are molecules that only differ in their three-dimensional arrangement of atoms. Stereoisomers can either be diastereomers or enantiomers. Diastereomers are non-mirror image molecules that only differ in the arrangement of their atoms in space. *Cis-trans* isomers are examples of diastereomers. Enantiomers are molecules that are nonsuperimposable mirror images. Enantiomers do not possess a symmetry element such as a plane of symmetry or point of symmetry. A molecule that is not superimposable on its mirror image is a chiral molecule. A molecule that is superimposable on its mirror image is achiral—it has a symmetry element that allows this.

One of the most common features of molecules that exist as enantiomeric pairs is a stereogenic center, commonly called a chiral center or a chiral C atom. Such a C atom has four different groups attached. For example, 2-bromobutane is a chiral molecule.

Br
|
*

H

2-bromobutane

The second C atom is bonded to a methyl group, ethyl group, hydrogen atom, and bromine atom. Thus, this C atom is a stereogenic center. A common way to show enantiomers is to use a Fischer projection. The stereogenic center is the intersection of two perpendicular lines. The groups bonded to the horizontal line come toward the viewer out of the plane of the paper and the groups bonded to the vertical line are located behind the plane of the paper. The Fischer projections for both enantiomers of 2-bromobutane are as follows:

Br   Br
| |
CH_3CH_2———CH_3 CH_3———CH_2CH_3
| |
H   H

The most common physical properties of enantiomers are essentially the same. They differ in their interaction with plane-polarized light. One enantiomer rotates the plane of polarized light in one direction and the other rotates the plane-polarized light the same magnitude but in the opposite direction. The enantiomer that rotates this light in the clockwise direction is called dextrorotary (to the right). The enantiomer that rotates the light in the counterclockwise direction is called levorotary (to the left). A plus sign in parentheses, (+), is used to designate dextrorotary, and a minus sign in parentheses, (–), is used to designate levorotary. Enantiomers are optically active because of their interaction with plane-polarized light. A 50-50% mixture of enantiomers, a racemic mixture, is not optically active because one cancels the optical activity of the other. Achiral molecules are also optically inactive. Optical activity is measured in a polarimeter.

Chemists are concerned with the configuration, i.e., spatial arrangement, of the groups bonded to the chiral center. The configuration at the stereogenic center is exactly opposite when considering enantiomers. The actual configuration of groups bonded to the stereogenic center is called the absolute configuration. The absolute configuration is not related to the direction in which an enantiomer rotates the plane of polarized light. To designate which configuration is at the stereogenic center a set of rules are used. They are called the Cahn-Ingold-Prelog rules. Following these rules tells if the absolute configuration is either R (rectus, the right-handed molecule) or S (sinister, the left-handed molecule). In general, the four groups are ranked from highest (1) to lowest priority (4). Higher priority groups have higher atomic numbers than lower priority groups. After ranking, the lowest priority group is rotated to the back away from the viewer and the remaining three groups are observed. If the line that connects the highest to next two lower priority groups is clockwise, then it is assigned the R configuration, and if the line that connects these groups is counterclockwise, then it is assigned the S configuration. For example, consider (R)-2-bromobutane.

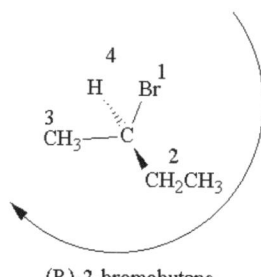

(R)-2-bromobutane

Scientists sometimes assign the relative configuration, which is the experimentally determined relationship between the configuration of a molecule to a reference molecule. Historically, the first reference molecule was glyceraldehyde, $HOCH_2CH(OH)CHO$. The (+) and (–) enantiomers were arbitrarily assigned D and L, respectively. To determine the relative configuration of other molecules, if possible, they were converted to glyceraldehyde.

Some molecules have more than one stereogenic center. A maximum of 2^n stereoisomers can exist where n is the number of stereogenic centers. For example, if a molecule has two stereogenic centers then a maximum of four (2^2) stereoisomers exist. The absolute configurations of molecules with two chiral centers are (R, R), (R, S), (S, S), and (S, R). The (R, R) and (S, S) molecules are enantiomers, and the (R, S) and (S, R) are also enantiomers. All other combinations of these molecules are diastereomers. Less than the maximum number of stereoisomers is obtained if one of the molecules has a plane, point, or axis of symmetry. A molecule with two or more stereogenic centers and a plane of symmetry is called a *meso* compound.

Stereochemistry may be used to monitor organic reaction mechanisms. If a reaction takes place in which the stereogenic center is not changed, then there is retention of configuration; the same configuration is found in the product as the reactant. If the stereogenic center is attacked, the configuration may be retained, inverted, or racemized. Retention of configuration means that the attacking group occupies the same position as the leaving group did. Inversion of configuration means the attacking group occupies the position opposite to the leaving group. Racemization means that equal amounts of both configurations are produced, yielding a racemic mixture.

Test Yourself

1. Draw the structure of *(S)*-2-iodohexane.

2. Write the complete name for the following molecule.

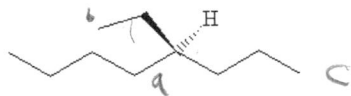

3. Define the following terms:

   a. chiral

   b. achiral

4. List two types of molecular symmetries that guarantee molecules will be achiral.

5. Draw the structure of 3,4-dimethyl-1-pentene and identify the stereogenic center.

6. Write a paragraph that describes how a polarimeter can be used to measure optical activity.

7. A solution contains a 60-40% mixture of enantiomers. What is the optical purity of this mixture?

8. What is the optical purity of a racemic mixture?

9. A solution of enantiomerically pure 2-butanol contains 13.3 g of 2-butanol dissolved in 150 mL of solution. This solution has an observed rotation of 1.2° when the length of the polarimeter tube is 10 cm. Calculate the specific rotation of 2-butanol.

10. Explain the difference between the absolute and relative configuration of a chiral molecule.

11. a. How many different stereoisomers of 2,3-dichlorohexane exist?

    b. Write the names of each.

12. What are the stereochemical relationships of all stereoisomers of 2,3-dichlorohexane (Problem 11)?

13. An optically pure enantiomer has a rotation of +6.8°. Calculate the optical purity of a mixture of enantiomers that has an observed rotation of 1.9°.

14. Describe the relationship of the following molecules.

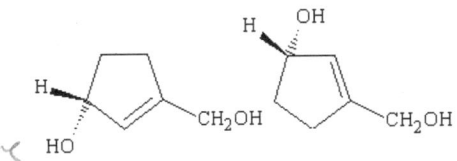

15. Draw all isomers of tribromocyclopropane and indicate which ones exist as enantiomeric pairs.

16. A molecule has three stereogenic C atoms.

    a. What is the maximum number of stereoisomers that can exist?

    b. How would these stereoisomers be designated?

17. Draw the Fischer projections for both stereoisomers of 1,2-diiodopropane.

18. a. What product(s) results when 1-hexene reacts with HCl?

    b. Describe the stereochemistry of this reaction.

19. Completely describe the stereochemistry of the reaction of (Z)-2-hexene and Br_2 in CCl_4.

20. Diastereomers are

    a. stereoisomers that are nonsuperimposable non-mirror images.

    b. stereoisomers that are nonsuperimposable mirror images.

    c. isomers that only differ by rotations around single bonds.

    d. isomers that only differ in the bonding arrangement of the atoms.

21. An optically pure substance is

    a. optically inactive because it is composed of a 50:50 mixture of enantiomers.

    b. an optically inactive enantiomer.

    c. optically active because it is composed of only one enantiomer.

    d. an optically active racemic mixture.

22. Which of the following groups has the highest priority using the Cahn-Ingold-Prelog rules for assigning absolute configuration?

    a. $-CH=CH_2$

    b. $-C\equiv CH$

    c. $-C_6H_{13}$

    d. $-C\equiv N$

23. Consider the following structures.

    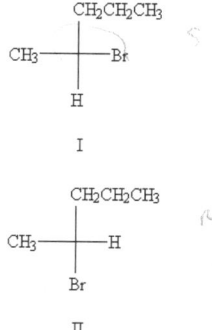

    Which of the following best describes the relationship of structures I and II?

    a. structural isomers

    b. diastereomers

    c. enantiomers

    d. no relationship between these structures

24. *Cis-trans* isomers can best be described as

    a. diastereomers

    b. enantiomers

    c. stereoisomers

    d. a and b

    e. a and c

25. Consider the following structure.

    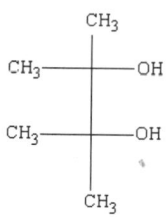

    Which of the following best describes this molecule?

    a. This molecule has two stereogenic centers and is optically active.

    b. This molecule has two stereogenic centers but the molecule is achiral.

    c. This molecule has one stereogenic center but the molecule is achiral.

    d. None of these are correct.

26. Which of the following physical properties differ when comparing (R)-2-hexanol and (S)-2-hexanol?

    a. refractive index

    b. normal boiling point

    c. density

    d. all are the same

27. Which of the following best describes the product of the reaction of (E)-2-butene and hydrogen chloride?

    a. The product is not optically active and contains no stereogenic C atom.

    b. The product is optically active but contains no stereogenic C atom.

    c. The product is not optically active and contains one stereogenic C atom.

    d. The product is optically active and contains a stereogenic C atom.

28. Which of the following is a correct statement regarding any *S*-enantiomer?

    a. It is levorotary.

    b. It is dextrorotary.

    c. It is a mixture of (+) and (−) isomers.

    d. It is the mirror image of the *R*-enantiomer.

29. Which of the following molecules have stereogenic C atoms?

    a. 1,2-dichlorobutane

    b. 1-chloro-2-methylcyclopentane

    c. 1,1,2-trichloropropane

    d. all of the above

30. Consider the following two molecules.

    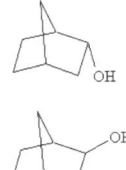

    Which of following best describes the relationship of these two molecules?

    a. diastereomers

    b. enantiomers

    c. conformational isomers

    d. no stereochemical relationship

31. Consider the following compound.

    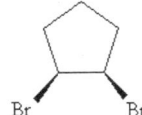

    Which of the following best describes this compound?

    a. It is achiral.

    b. It contains two stereogenic C atoms.

    c. It is a meso structure.

    d. all of the above

32. Which of the following would best describe one way to resolve a pair of enantiomeric acids?

    a. Enantiomers cannot be resolved.

    b. First, convert the enantiomers to salts using a chiral enantiomerically pure base. Then, use fractional crystallization to separate the diastereomeric salts.

    c. First, convert the enantiomers to salts using a chiral base. Then, use fractional distillation to separate the diastereomeric salts.

    d. First, convert the enantiomers to salts using an achiral base. Then, use fractional crystallization to separate the diastereomeric salts.

33. Which of the following occurs in a reaction that undergoes racemization?

    a. a reaction in which the groups on the stereogenic C atom of the product are the opposite configuration to that of the reactant

    b. a reaction in which the groups on the stereogenic C atom of the product have the same arrangement as they did in the reactant

    c. a reaction in which an optically active reactant produces products that are optically inactive

    d. a reaction in which an optically inactive reactant produces products that are optically active

34. Consider the following molecule.

    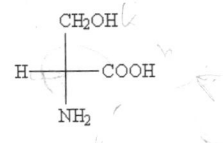

    Which of the following is correct concerning this molecule?

    a. It has *R* configuration

    b. It has *S* configuration.

    c. It represents a racemic mixture.

    d. It has neither R nor *S* configuration.

35. Which of the following produces two diastereomers when $Br_2(CCl_4)$ is added to the molecule?

    a. methylenecyclopentene

    b. 1-methylcyclopentene

    c. 3-methylcyclopentene

    d. 4-methylcyclopentene

✓ Check Yourself

1. The structure of (S)-2-iodohexane is

    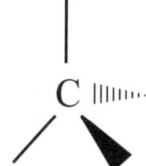

    (Stereochemistry nomenclature)

2. The name of this molecule is (S)-4-ethyloctane because an ethyl group is bonded to the fourth C atom in a chain of eight C atoms. It is an S configuration because the three highest-priority groups are arranged counterclockwise. **(Stereochemistry nomenclature)**

3. a. Molecules are chiral if they are nonsuperimposable mirror images of each other.

    b. Molecules are achiral if their mirror images are superimposable. The word "chiral" comes from the Greek word *cheiros*, which means "hand." **(Enantiomers)**

4. Two types of symmetry found in achiral molecules are:

    a. plane of symmetry

    b. point of symmetry

    (Symmetry)

5. The structure of 3,4-dimethyl-1-pentene is as follows with the stereogenic C atom designated with an asterisk. **(Stereogenic C atoms)**

    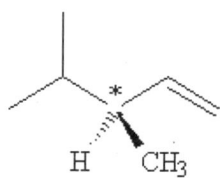

    3,4-dimethyl-1-pentene

6. A polarimeter has a source of light that passes through a polarizing filter; e.g., a nicol prism. The resulting polarized light usually passes through a 1-dm polarimeter tube that contains the substance that is being measured. On the other side of this tube is an analyzing polarizing filter that can be rotated in either direction. The number of degrees and the direction that the filter must be rotated is the observed rotation, α, of the sample. **(Measuring optical activity)**

7. Optical purity or percent enantiomeric excess is calculated by subtracting the percent of one enantiomer from that of the other. Thus, the optical purity of this mixture is 60% – 40% or 20%. **(Optical purity)**

8. A racemic mixture is a 50-50% mixture of enantiomers; hence, its optical purity is 0% (optical purity = 50% - 50%) **(Optical purity)**

9. Specific rotation, [a], is calculated as follows:

    $$|\alpha| = \frac{\alpha}{c \times l}$$

In this equation, l is the observed rotation, $1.2°$, c is the concentration in grams of dissolved substance per mL of solution, 13.3 g/150 mL = 0.090 g/mL, and l is the length of the polarimeter tube in decimeters, 10 cm = 1.0 dm. Substituting the values into the equation yields the following.

$[\alpha] = \alpha/(c \times l) = 1.2°/(0.090 \text{ g/mL} \times 1.0 \text{ dm}) = 14°$ **(Specific rotation)**

10. The absolute configuration is the exact arrangement in space of the substituent groups bonded to the stereogenic C atom. The relative configuration is the experimentally determined relationship between the configurations of two compounds. Relative configurations are typically found through chemical reactions that do not involve the stereogenic center. Relative configurations can be obtained without knowledge of the absolute configuration. **(Configurations)**

11. Four different stereoisomers of 2,3-dichlorohexane exist: (2*R*, 3*R*), (2*R*, 3*S*), (2*S*, 3*S*), and (2*S*, 3*R*). **(Stereoisomers)**

12. (2*R*, 3*R*)-dichlorohexane and (2*S*, 3*S*)-dichlorohexane are enantiomers. (2*R*, 3*S*)-dichlorohexane and (2*S*, 3*R*)-dichlorohexane are enantiomers. All other combinations are diastereomers. **(Stereoisomers)**

13. To calculate the optical purity, divide the observed rotation by the rotation of the pure enantiomer and multiply by 100. **(Optical purity)**

Optical purity = $(1.9°/6.8°) \times 100 = 28\%$

14. These two molecules are enantiomers. First, rotate the second structure clockwise so that the two CH_2OH groups are next to each other, and then observe that the molecules are mirror images and are not superimposable. **(Stereoisomers)**

15. Both *cis* and *trans* 1,2,3-tribromocyclopropane are achiral because of their planes of symmetry.

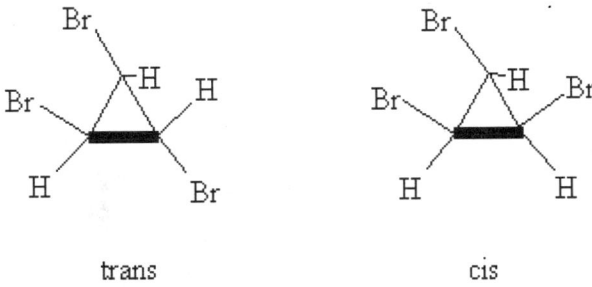

    trans   cis

However, 1,1,2-tribromocyclopropane exists as an enantiomeric pair.

(Stereoisomers/meso structures)

16. a. The maximum number of stereoisomers is calculated by raising 2 to the n power, 2^n, where n is the number of chiral centers. Hence, there can be eight isomers, 2^3.

    b. The following shows how these stereoisomers would be designated.

Isomer	Stereogenic C centers	Enantiomer	Stereogenic C centers
1	R R R	5	S S S
2	R R S	6	S S R
3	R S R	7	S R S
4	S R R	8	R S S

17. The Fischer projections for 1,2-diiodopropane are as follows.

(Fischer projections)

18. a. The product of this reaction is 2-chlorohexane because the reaction follows Markovnikov's addition across the double bond.

    b. Because the intermediate secondary carbocation can be attacked equally by the chloride ion from above or below the plane of the cation, a racemic mixture of (R)- and (S)-2-chlorohexane results. **(Stereochemistry in reactions)**

19. Two enantiomers, (2R, 3R)-2,3-dibromohexane and (2S, 3S)-2,3-dibromohexane are produced because of the anti addition of the Br atoms.

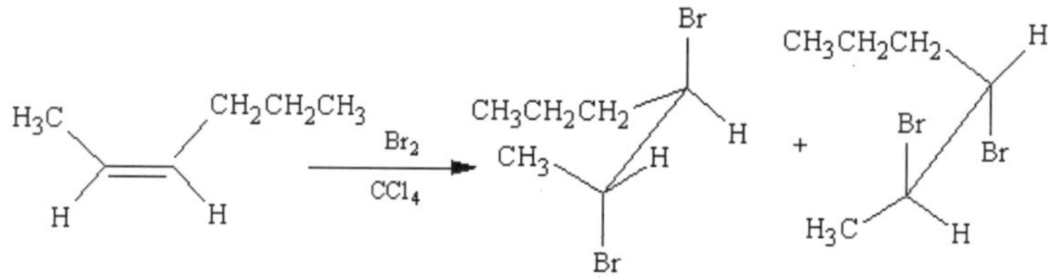

20. a. Diastereomers are stereoisomers that are nonsuperimposable nonmirror images. *Cis-trans* isomers are examples of diastereomers. **(Diastereomers)**

21. c. An optically pure substance is optically active because it is composed of only one enantiomer. It is 100% of either the R or S form of an enantiomer. **(Optical purity)**

22. d. The nitrile group, $-C \equiv N$, has the highest priority because the C is considered to be bonded to three N atoms. **(Priority rules)**

23. c. Structures I and II are enantiomers because they are nonsuperimposable mirror-image molecules. **(Enantiomers)**

24. e. *Cis-trans* isomers, geometric isomers, are both stereoisomers and diastereomers. **(Diastereomers)**

25. b. This molecule has two chiral centers, but the molecule is achiral because it is a meso structure. **(Stereoisomers with more than one chiral center)**

26. d. The most common physical properties of enantiomers, such as refractive indices, boiling points, and densities, are the same except for their specific rotations. **(Enantiomers)**

27. c. The product, 2-chlorobutane, is not optically active because a racemic mixture results. The second C atom in 2-chlorobutane is a stereogenic center because it has four different groups attached. **(Stereochemistry in reactions)**

28. d. It is the mirror image of the *R*-enantiomer. **(Enantiomers)**

29. d. All of the molecules listed, a, b, and c, have chiral C atoms. **(Stereogenic C atoms)**

30. a. These molecules are diastereomers because they are non-mirror-image molecules that only differ in the arrangement of their atoms in space. **(Diastereomers)**

31. d. All of the above are correct. This molecule, *cis*-1,2-dibromocyclobutane, has two stereogenic centers but also has a plane of symmetry, which makes the mirror image superimposable. This means it is a meso compound. **(Stereoisomers/meso structures)**

32. b. First, convert the enantiomers to salts using a chiral base such as an alkaloid. Then use fractional crystallization to separate the diastereomeric salts. This can be accomplished because diastereomers have different solubilities. **(Resolving enantiomers)**

33. c. A reaction in which an optically active reactant produces products that are optically inactive. **(Stereochemistry in reactions)**

34. a. It is the *R* configuration because the three highest priority groups are arranged in a clockwise direction. **(Enantiomers)**

35. c. 3-Methylcyclopentene produces the following two diastereomers.

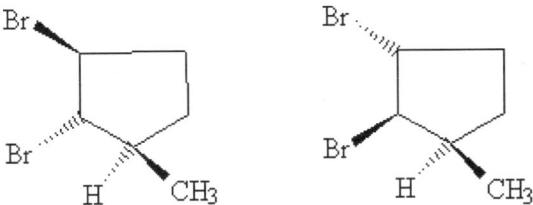

Methylenecyclopentene only produces one product. Both 1-methylcyclopentene and 4-methylcyclopentene produce enantiomeric products. **(Stereochemistry in reactions)**

Grade Yourself

Circle the number of questions you missed, then fill in the total incorrect for each topic. If you answered more than three questions incorrectly, you need to focus on that topic. (If a topic has less than three questions and you had at least one wrong, we suggest you study that topic also. Read your textbook, a review book, or ask your teacher for help.)

Subject: Stereochemistry

Topic	Question Numbers	Number Incorrect
Stereochemistry nomenclature	1, 2	
Enantiomers	3, 23, 26, 28, 34	
Symmetry	4	
Stereogenic C atoms	5, 29	
Measuring optical activity	6	
Optical purity	7, 8, 13, 21	
Specific rotation	9	
Configurations	10	
Stereoisomers	11, 12	
Stereoisomers/meso structures	15, 31	
Fischer projections	17	
Diastereomers	20, 24	
Priority rules	22	
Stereoisomers with more than one chiral center	25	
Stereochemistry in reactions	27, 33, 35	
Resolving enantiomers	32	

Nucleophilic Substitution Reactions

7

Brief Yourself

The general equation for nucleophilic substitution reactions is

$$\text{Nu:}^- + \text{R–L} \rightarrow \text{R–Nu} + \text{L:}^-$$

in which Nu:⁻ is a nucleophile, R–L is the substrate molecule attacked by the nucleophile, R–Nu is the substituted product, and L:⁻ is the leaving group. Nucleophiles are Lewis bases (electron-pair donors) that are either anions (e.g., X^-, OH^-, OR^-, CN^-, or N_3^-) or neutral molecules (e.g., NH_3, H_2O, CH_3OH, or CH_3CH_2OH). Common substrates, R–L, in nucleophilic substitution reactions are alkyl halides, RX, and sulfonates, $ROSO_2R'$.

The two principal mechanisms that describe nucleophilic substitution reactions are substitution nucleophilic bimolecular, S_N2, and substitution nucleophilic unimolecular, S_N1. The S_N2 mechanism is concerted and follows second-order kinetics.

$$\text{Rate} = k\,[\text{R–L}][\text{Nu:}^-]$$

The S_N1 mechanism goes through a carbocation intermediate and follows first-order kinetics.

$$\text{Rate} = k\,[\text{R–L}]$$

In the S_N2 reaction the nucleophile attacks the substrate from the direction opposite to the leaving group. This is sometimes called "backside attack." During the reaction, the nucleophile bonds to the C atom as the leaving group is displaced. Inversion of configuration characterizes the S_N2 reactions that take place at stereogenic centers. The order of reactivity for S_N2 reactions is as follows.

$$CH_3 > 1^o > 2^o > 3^o$$

This is explained in terms of increasing steric hindrance of the attacking nucleophile going from methyl to tertiary substrates. The order of S_N2 reactivity of alkyl halides is as follows.

$$RI > RBr > RCl > RF$$

This is explained in terms of bond energies. The C–I bond energy is lowest (the weakest bond), which makes it easier to break than the other carbon-halogen bonds. Also the more diffuse iodide ion can better carry away the negative charge. The C–F bond is the strongest and hardest to break, making alkyl fluorides the least reactive in S_N2 reactions. S_N2 reactions take place most readily in polar aprotic solvents.

The S_N1 reaction takes place in two steps. First the leaving group departs, producing a carbocation. This is the rate determining step. To complete the mechanism, the nucleophile bonds to the carbocation.

$$R–L \rightarrow R^+ + L^- \text{ (rate-determining step)}$$

$$R^+ + Nu:^- \rightarrow R–Nu$$

Because the formation of a carbocation is the rate-determining step, carbocation stability determines the order of substrate reactivity.

$$3^\circ > 2^\circ > 1^\circ > CH_3$$

Partial racemization is found in S_N1 reactions because the trigonal planar geometry of the carbocation intermediate allows the nucleophile to attack from two directions. Complete racemization is generally not found because slightly more inversion than retention of configuration occurs. The order of S_N1 reactivity of alkyl halides is as follows.

$$RI > RBr > RCl > RF$$

The rates of S_N1 reactions increase in more polar solvents because they help to better solvate the intermediate ions. The strength and concentration of the nucleophile do not affect the rate of S_N1 reactions because the nucleophile is not in the rate-determining step. S_N1 intermediates sometimes undergo rearrangements to produce a more stable carbocation.

Elimination reactions, E1 and E2, compete with substitution reactions. In the E2 mechanism a base (which is also a nucleophile) abstracts a proton and at the same time a double bond forms and the leaving group departs. The E2 mechanism follows second-order kinetics. The rate of E2 reactions depends on the leaving group location as follows.

$$3^\circ > 2^\circ > 1^\circ$$

E2 is favored over E1 using strong bases. In the E1 mechanism, the leaving groups break free in the rate-determining step, producing a carbocation. Then a proton is removed by the base from an adjacent C atom resulting in the formation of a double bond.

Test Yourself

1. a. Write the general equation for any nucleophilic substitution reaction.

    b. Write a specific equation that shows the reaction of an alkyl halide and hydroxide.

2. What is the organic product of the reaction of potassium fluoride and l-bromobutane?

3. a. What is the order of reactivity of alkyl halides to nucleophilic substitution?

    b. Explain this ranking.

    (Consider the S_N2 reaction of OH^- and methyl chloride for Problems 4 to 7.)

    $$OH^- + CH_3Cl \rightarrow CH_3OH + Cl^-$$

4. Write the rate expression for this reaction. Explain the meaning of this equation.

5. What does the 2 in S_N2 mean in this reaction mechanism?

6. Draw the structure of the activated complex for this reaction. Is this activated complex polar or nonpolar? Explain.

7. Describe the energy pathway for this S_N2 reaction in solution. *grou*

8. Consider the S_N2 displacement when hydroxide reacts with (R)-2-bromobutane to produce 2-butanol.

    a. Describe the stereochemistry of the reaction and the absolute configuration of the product.

    b. What products result if racemic 2-bromobutane is used in place of (R)-2-bromobutane?

    not cover d

    (Consider the S_N1 reaction of boiling methanol, CH_3OH, and *t*-butyl bromide for Problems 9 to 12.)

    $$CH_3OH + (CH_3)_3CBr \rightarrow (CH_3)_3COCH_3 + HBr$$

9. Write the mechanism for this reaction and indicate the rate-determining step.

10. Write the rate equation for this reaction. Explain.

11. Explain why you would not expect this reaction to undergo an S_N2 mechanism.

12. Draw the structure of the activated complex for the rate-determining step of this reaction.

13. Consider the S_N1 displacement when water reacts with (R)-3-bromo-3-methylhexane. Describe the stereochemistry of the reaction and the absolute configuration of the product.

14. What is the principal product that results when 3-bromo-2-methylpentane reacts with water? Explain your reasoning.

15. What two products result when 1-chloro-2-methylpropane reacts with sodium ethoxide?

16. a. Draw the structure of ethyl *p*-toluenesulfonate.

    b. What is the importance of using alkyl sulfonate esters in nucleophilic substitution reactions?

17. a. Write the equation for the reaction of sodium hydrogen sulfide with *n*-hexyl *p*-toluene sulfonate.

    b. What mechanism does it follow?

18. Starting with *t*-butyl chloride show a synthesis that gives a reasonable yield of 1-cyano-2-methylpropane, $(CH_3)_2CHCH_2CN$.

19. One of the products of the solvolysis of bromomethylcyclopentane in methanol is methyl cyclohexyl ether. Write a complete mechanism that accounts for the formation of this product.

20. Which of the following is most reactive in an S_N2 reaction?

    a. CH_3Br

    b. CH_3CH_2Br

    c. $(CH_3)_2CHBr$

    d. $(CH_3)_3CBr$

21. Which of the following is the strongest nucleophile?

    a. I^-

    b. F^-

    c. CH_3OH

    d. H_2O

22. Which of the following most readily undergoes S_N1 displacements?

    a. 1-chloropentane

    b. 2-chloropentane

    c. 2-chloro-2-methylbutane

    d. 1-chloro-2-methylbutane

23. Which is the best solvent to use for the solvolysis reaction of *t*-butyl chloride?

    a. acetic acid

    b. hexane

    c. carbon tetrachloride

    d. water

24. What is the major product(s) of the reaction of 2-bromopentane with sodium ethoxide in ethanol?

    a. 2-ethoxypentane

    b. *cis*-2-pentene

    c. *trans*-2-pentene

    d. b and c

25. Which of the following is **not** true about S_N2 reactions?

    a. They follow a two-step bimolecular mechanism

    b. The rates of the reactions depend on both the substrate and nucleophile concentrations.

    c. They do not undergo rearrangements.

    d. Methyl and primary substrates react faster than secondary and tertiary substrates.

26. Which best describes the reaction of (*S*)-2-bromohexane with a water-ethanol mixture to produce 2-hexanol?

    a. complete retention of configuration

    b. complete inversion of configuration

    c. complete racemization

    d. partial racemization and partial inversion.

27. Consider the following alkyl halide.

    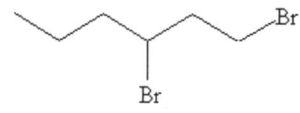

    1,3-dibromohexane

    If one mole of this dibromide is mixed with one mole of NaI in acetone, what is the principal product of the reaction?

    a. 1-bromo-3-iodohexane

    b. 3-bromo-1-iodohexane

    c. 1,3-diiodohexane

    d. none of these

28. Which of the follow mechanisms is favored when a weak base in a polar solvent is reacted with a tertiary substrate that has a good leaving group?

    a. S_N1

    b. S_N2

    c. E1

    d. E2

    e. a and c

    f. b and d

29. Which of the following bases would give the best yield for the substitution product when reacted with 2-bromopropane?

    a. OH^-

    b. CH_3COO^-

    c. $CH_3CH_2O^-$

    d. NH_2^-

30. Which of the following would be most reactive under S_N1 conditions?

    a. $(CH_3)_3C–Br$

    b. $CH_3CH_2CH_2CH_2–Br$

    c. $(CH_3)_2CH–Br$

    d. $(CH_3)_3CCH_2–Br$

31. Which of the following is the best nucleophile in an S_N2 reaction where competition from E2 is unimportant?

    a. H_2O

    b. Cl^-

    c. CH_3COO^-

    d. OH^-

32. Consider the substitution reaction that occurs when ethyl bromide reacts with potassium *t*-butoxide. What happens to the reaction rate when both the ethyl bromide and potassium *t*-butoxide concentrations are doubled?

    a. The reaction rate does not change.

    b. The reaction rate is twice the initial rate.

    c. The reaction rate is four times the initial rate.

    d. The reaction rate is one-half of the initial rate.

33. Which of the following reacts under S_N2 conditions with sodium cyanide, NaCN, to produce the following product?

a. *cis*-1-iodo-2-methylcyclopentane

b. *trans*-1-iodo-2-methylcyclopentane

c. *cis*-2-methylcyclopentanol

d. *trans*-2-methylcyclopentanol

34. What is the major product when 3-bromo-2, 2-dimethylbutane undergoes an E1 reaction?

    a. 3,3-dimethyl-1-butene

    b. 2,3-dimethyl-2-butene

    c. 2,3-dimethyl-1-butene

    d. none of these

35. Usually primary halides are least reactive in S_N1 solvolysis reactions. Nonetheless, $CH_3CH_2-O-CH_2-Cl$, a primary halide, rapidly undergoes solvolysis reactions. Which of the following best explains this apparent anomaly?

    a. Cl is a much better leaving group in this molecule because it forms a resonance-stabilized carbocation.

    b. The O atom inductively withdraws electron density from the C atom that is bonded to the Cl atom, making the Cl atom a better leaving group.

    c. This molecule forms an allylic carbocation that is more stable than tertiary carbocations.

    d. none of these

✔ Check Yourself

1. a. $Nu:^- + R–L \rightarrow R–Nu + L:^-$
   b. $OH^- + CH_3–Br \rightarrow CH_3–OH + Br^-$ (**Nucleophilic substitution reactions**)

2. The reaction of KF and $CH_3CH_2CH_2CH_2–Br$ produces $CH_3CH_2CH_2CH_2–F$. (**Nucleophilic substitution reactions**)

3. a. RI > RBr > RCl >> RF
   b. Alkyl iodides are most reactive in nucleophilic substitution reactions because I is a better leaving group. Alkyl fluorides are unreactive in most nucleophilic substitution reactions because F is a poor leaving group which results from the strong C–F bond. (**Substrate reactivity**)

4. Rate = k $[CH_3Cl][OH^-]$
   The reaction is first order with respect to both reactants and is second order overall. This means that if either reactant concentration is doubled the rate of the reaction doubles, and if both reactant concentrations are doubled the rate of the reaction increases four times. (**S_N2 kinetics**)

5. The "2" in S_N2 means bimolecular. Thus, two particles (CH_3Cl and OH^-) are in the activated complex of the rate-determining step of the reaction mechanism. (**S_N2 kinetics**)

6. The activated complex for this reaction is as follows.

    $$[\overset{\delta-}{HO--}CH_3\overset{\delta-}{--Cl}]$$

    The activated complex is essentially nonpolar because the partial negative charge that develops on the leaving group, Cl, is canceled by the partial negative charge that develops on the OH group. (**S_N2 kinetics**)

7. The OH^- attacks the methyl chloride $180°$ from the Cl, the leaving group. As the OH^- approaches, the energy increases until the activated complex forms (see Problem 6). The energy required to produce the activated complex is the activation energy. The activated complex then breaks apart and produces the products, methanol and Cl^-. This is the energy profile of a bimolecular collision between the reactants in solution. (**S_N2 kinetics**)

8. a. In the S_N2 mechanism, the nucleophile attacks the molecule from the "backside" of the leaving group; hence, inversion of configuration is found in S_N2 displacements. In this reaction, (R)-2-bromobutane is attacked by the OH^-, producing (S)-2-butanol.
    b. Reaction of OH^- with racemic (R)-2-bromobutane will produce racemic 2-butanol. (**S_N2 stereochemistry**)

9. S_N1 reactions follow a two-step mechanism. First, the *t*-butyl bromide undergoes a unimolecular dissociation in which the Br leaves with the pair of electrons that bond it to the C atom. The products of this step are the *t*-butyl carbocation and the bromide ion. This is the rate-determining or slowest step in the mechanism.

    $(CH_3)_3C–Br \rightarrow (CH_3)_3C^+ + Br^+$

    In the second step, the carbocation bonds to CH_3OH, producing $(CH_3)_3C–O^+(H)CH_3$. This is a fast step.

    $(CH_3)_3C^+ + CH_3OH \rightarrow (CH_3)_3C–O^+(H)CH_3$

    In the final step of the mechanism, the proton is lost to the solvent methanol.

    $(CH_3)_3C–O^+(H)CH_3 + CH_3OH \rightarrow (CH_3)_3C–OCH_3 + CH_3OH_2^+$ (**S_N1 mechanism**)

10. Because the rate-determining step only involves the *t*-butyl bromide and not the methanol the rate equation for the reaction is as follows. (**S_N1 kinetics**)

    $Rate = k\ [(CH_3)_3CBr]$

11. In an S_N2 mechanism the nucleophile attacks the substrate on the opposite side of the leaving group. The three methyl groups in the *t*-butyl bromide block the approach of the hydroxide, preventing it from undergoing an S_N2 mechanism. (**S_N1 versus S_N2 mechanism**)

12. The activated complex for the rate-determining step results when the C–Br bond is partially broken. (**S_N1 kinetics**)

$$\left[\begin{array}{c} CH_3 \\ | {\scriptstyle\delta+} \quad {\scriptstyle\delta-} \\ CH_3-C----Br \\ | \\ CH_3 \end{array}\right]$$

13. In S_N1 reactions, the highest-energy intermediate is a carbocation, which is trigonal planar. Hence, the nucleophile, water, can attack from either above or below the plane of the carbocation, producing both *R* and *S* stereoisomers. In other words, racemization characterizes S_N1 reactions. In this reaction, (*R*)-3-bromo-3-methylhexane reacts with water to produce racemic 3-methyl-3-hexanol. Note that complete racemization does not always occur because of the association of the leaving group with the carbocation. (**S_N1 stereochemistry**)

14. The principal product that forms in this reaction is 2-methyl-2-pentanol, not 2-methyl-3-pentanol. An important feature of substrate molecules in S_N1 reactions is that they rearrange to form more stable carbocations. In this reaction, a hydride shift occurs forming the more stable tertiary carbocation, which is then attacked by the water molecule. (**S_N1 mechanism**)

Secondary carbocation   Tertiary carbocation

15. Sodium ethoxide, NaOCH₂CH₃, is a strong base because the ethoxide ion, CH₃CH₂O[−], is the conjugate base of the very weak acid ethanol. The ethoxide ion is also a good nucleophile. Thus, elimination and substitution reactions compete producing an elimination product, an alkene, and a substitution product, an ether. The two products are isobutene (2-methylpropene) and ethyl isobutyl ether (1-ethoxy-2-methylpropane). (**S_N2 versus E2 reactions**)

16.   a. The structure of ethyl *p*-toluenesulfonate, a tosylate, is as follows

Ethyl p-toluenesulfonate

b. Tosylates are good leaving groups in nucleophilic substitution reactions because this structure can resonance stabilize the negative charge that develops on the leaving group. (**Leaving groups**)

17.   a. SH[−] + CH₃CH₂CH₂CH₂CH₂CH₂–OTs → CH₃CH₂CH₂CH₂CH₂CH₂–SH + TsO[−]
b. This follows an S_N2 mechanism. (**S_N2 reactions**)

18.   (CH₃)₃C–Cl + NaOC(CH₃)₃ → (CH₃)₂C=CH₂
(CH₃)₂C=CH₂ + HBr(peroxides) → (CH₃)₂CHCH₂Br
(CH₃)₂CHCH₂Br + CN[−] → (CH₃)₂CHCH₂CN + Br[−] (**Multistep synthesis**)

19.

(S$_N$1 mechanism)

20. a. Methyl halides are significantly more reactive in S$_N$2 reactions because of less steric hindrance **(S$_N$2 reactions)**

21. a. I$^-$ is the strongest nucleophile because is carries a negative charge and it is a diffuse ion with 54 somewhat loosely held electrons. **(Nucleophilicity)**

22. c. 2-Chloro-2-methylbutane undergoes S$_N$1 reactions most readily because it is the only tertiary chloride listed. The others are primary and secondary chlorides. **(S$_N$1 reactivity)**

23. d. The solvolysis of *t*-butyl chloride follows an S$_N$1 mechanism. The transition state is polar because of the ionization of *t*-butyl chloride; thus, the best solvent for solvolysis is the most polar (has the highest dielectric constant), water. **(S$_N$1 mechanism)**

24. d. Secondary bromides more readily undergo elimination reactions with strong bases such as sodium ethoxide; thus, the major products are both *cis*- and *trans*-2-pentene. **(Substitution versus elimination)**

25. a. S$_N$2 reactions follow a one-step bimolecular mechanism. **(S$_N$2 reactions)**

26. d. Partial racemization and partial inversion best describes this reaction because water is a weak nucleophile. Thus, the reaction mainly follows the S$_N$1 mechanism which produces racemization, but some of the reactants follow the S$_N$2 pathway, which produces inversion. Also, association of the bromide with the carbocation does not allow for complete racemization. **(S$_N$1 versus S$_N$2)**

27. b. 3-Bromo-1-iodohexane is the principal product because the I$^-$ is a good nucleophile and will more readily attack the less hindered primary C atom. With only one mole of iodide available, the major product will be the monosubstituted product and not the disubstituted one. **(S$_N$2 reactions)**

28. e. S$_N$1 and E1 both share the listed properties of polar solvent and weak base with a tertiary substrate and good leaving group. **(S$_N$1 and E1)**

29. b. Acetate, CH_3COO^-, is a weak base and thus would not initiate an elimination reaction. All of the other bases listed are strong bases that yield high percents of the elimination product, propene. **(Substitution versus elimination)**

30. a. The tertiary bromide, *t*-butyl bromide, would be most reactive because tertiary carbocations are more stable than either secondary or primary carbocations. **(S$_N$1 reactivity)**

31. d. OH¯ is the best nucleophile because it is the strongest base listed. Nucleophilicity often parallels basicity. (**S$_N$2 reactions**)

32. c. The reaction rate is four times the initial rate because this is an S$_N$2 reaction with each reactant being first order. Thus, doubling both concentrations increases the reaction rate to four times that of the initial rate. (**S$_N$2 kinetics**)

33. a. To obtain the *trans* product *cis*-1-iodo-2-methylcyclopentane must be used as the substrate. The good nucleophile, CN¯, would do an S$_N$2 displacement of the iodide. (**S$_N$2 reactions**)

34. b. The product of this reaction is 2,3-dimethyl-2-butene because a rearrangement in the C skeleton would occur to produce the most stable carbocation. The major product would then be the elimination product that is most highly substituted. (**E1 reaction**)

35. a. Cl is a much better leaving group in this molecule because it forms the resonance-stabilized carbocation, $CH_3CH_2-O-CH_2^+$, which forms at a much lower energy than a primary chloride. (**Leaving groups**)

Grade Yourself

Circle the number of questions you missed, then fill in the total incorrect for each topic. If you answered more than three questions incorrectly, you need to focus on that topic. (If a topic has less than three questions and you had at least one wrong, we suggest you study that topic also. Read your textbook, a review book, or ask your teacher for help.)

Subject: Nucleophilic Substitution Reactions

Topic	Question Numbers	Number Incorrect
Nucleophilic substitution reactions	1, 2	
Substrate reactivity	3	
S_N2 kinetics	4, 5, 6, 7, 32	
S_N2 stereochemistry	8	
S_N1 mechanism	9, 14, 19, 23	
S_N1 kinetics	10, 12	
Leaving groups	16, 35	
S_N1 versus S_N2 mechanism	11, 26	
S_N2 versus E2 reactions	15	
S_N1 stereochemistry	13	
S_N2 reactions	17, 20, 25, 27, 31, 33	
Multistep synthesis	18	
Nucleophilicity	21	
Substitution versus elimination	24, 29	
S_N1 and E1	28	
S_N1 reactivity	22, 30	
E1 reaction	34	

Alkynes

8

Brief Yourself

Alkynes are hydrocarbons that contain a C–C triple bond. The general formula for alkynes is as follows.

$$R–C\equiv C–R(H)$$

The general formula for the alkynes is C_nH_{2n-2}. The simplest alkyne is acetylene, $HC\equiv CH$. Acetylene is an important industrial chemical.

To write the IUPAC name of an alkyne follow the same general rules as alkenes but use *yne* as the ending. For example the IUPAC name for the following alkyne is 3,4-dimethyl-1-heptyne.

$$CH_3—(CH_2)_2—\overset{\displaystyle CH_3}{\overset{\displaystyle |}{CH}}—\overset{\displaystyle CH_3}{\overset{\displaystyle |}{CH}}—C\equiv CH$$

3,4-dimethyl-1-heptyne

The C–C triple bond is composed of one sigma and two pi bonds. The two pi bonds are perpendicular to each other. The two C atoms that make up a triple bond are *sp* hybridized; thus, each C atom has 50% *s* character. The geometry about the C–C triple bond is linear (180° bond angle). As a result of the linear geometry, stereoisomerism is not possible at these C atoms.

When a H atom is bonded to a C–C triple bond, it is significantly more acidic than H atoms bonded to either sp^3 or sp^2 C atoms. Strong bases such as sodium amide, $NaNH_2$, can withdraw the proton bonded to a C–C triple bond, producing an acetylide. Acid strength is related to the ability of the resulting conjugate base to hold the negative charge. Because *sp* hybridized C atoms can more readily accept the negative charge than sp^3 or sp^2 hybridized C atoms (due to their higher s character), terminal alkynes are more acidic than alkenes or alkanes.

The physical properties of alkynes resemble those of alkanes and alkenes. They are composed of nonpolar molecules that are soluble in nonpolar organic solvents and insoluble in polar solvents.

They have relatively low boiling points, but alkynes have higher boiling points than alkanes and alkenes with similar C skeletons and molecular masses.

Two principal reactions are used to prepare alkynes. One way is the nucleophilic substitution of a halide in a methyl or primary alkyl halide by a terminal acetylide.

$$RC \equiv C^- + H_3C-X \rightarrow RC \equiv CCH_3 + X^-$$

Alkynes can also be prepared by the double dehydrohalogenation of either geminal or vicinal dihalides.

$$RCH_2CX_2R' + 2NaNH_2 \rightarrow RC \equiv CR' + 2NaX$$

$$RCHXCHXR' + 2NaNH_2 \rightarrow RC \equiv CR' + 2NaX$$

Alkynes undergo many of the same reactions as alkenes. As a result of having two pi bonds, one mole of alkyne reacts with two moles of reagent. For example, alkynes undergo catalytic hydrogenation as follows.

$$RC \equiv CR' + 2H_2 \text{ (Pt cat.)} \rightarrow RCH_2CH_2R'$$

If Lindlar's catalyst is used, the hydrogenation produces a *cis*-alkene. A *trans*-alkene can be produced using Na in liquid ammonia, $NH_3(l)$. Alkynes react with two moles of hydrogen halides to produce geminal dihalides. This reaction follows Markovnikov's rule.

$$RC \equiv CH + HX \rightarrow RCX=CH_2 + HX \rightarrow RCX_2CH_3$$

In a similar manner, alkynes react with two moles of halogen, X_2, to produce tetrahalides.

$$RC \equiv CR' + X_2 \rightarrow RCX=CXR' \text{ } (trans) + X_2 \rightarrow RCX_2CX_2R'$$

When alkynes undergo acid-catalyzed hydration reactions they produce ketones. Initially, an enol results that rapidly tautomerizes to a ketone. This reaction is usually carried out in aqueous H_2SO_4 using Hg^{2+} as a catalyst.

$$RC \equiv CH + H_2O \text{ } (H^+, Hg^{2+}) \rightarrow [RC(OH)=CH_2] \rightarrow RCOCH_3$$

Alkynes oxidize in an aqueous neutral $KMnO_4$ solution to diketones.

$$RC \equiv CR' + KMnO_4(aq) \rightarrow RCOCOR'$$

Using more concentrated alkaline $KMnO_4$ solution, alkynes oxidatively cleave to produce two carboxylic acids after acid hydrolysis of the product.

$$RC\equiv CR' + KMnO_4(OH^-) \rightarrow RCOOH + R'COOH$$

The same products result when alkynes are oxidatively cleaved using ozone.

Terminal alkynes, ones with a triple bond at the end of a chain, react with silver(I) and copper(I) in ammonia to produce silver and copper acetylides.

$$RC\equiv CH + Ag^+ (NH_3) \rightarrow RC\equiv CAg + H^+$$

$$RC\equiv CH + Cu^+ (NH_3) \rightarrow RC\equiv CCu + H^+$$

Test Yourself

1. Describe the structure of C–C triple bond.

2. Compare the bond distance and bond angles in C–C single, double, and triple bonds.

3. Compare the s character of C atoms in single, double, and triple bonds.

4. Write the name of the following alkyne.
   $$CH_3 CH_2 C\equiv CC(CH_3)_3$$

5. Draw the structure of 4-chloro-4-methyl-2-heptyne.

6. What are the principal physical properties of alkynes?

7. Compare and explain the trend in acidity of acetylene, ethylene, and ethane.

8. a. Knowing that the amide ion, NH_2^-, is a stronger base than the acetylide ion, write an equation that shows the equilibrium established when amide ions react with acetylene.
   b. In what direction does this equilibrium lie?

9. Write an equation that shows how 1-butyne can be prepared using sodium acetylide, Na^+ $HC\equiv C^-$. Explain what happens in this reaction.

10. Using only methyl bromide, sodium amide, and sodium acetylide, show how 2-butyne may be synthesized.

11. Starting with 2-propanol and any necessary reagents show how propyne can be synthesized.

12. Write the equation for the hydration of 1-hexyne using H_2SO_4 and $HgSO_4$.

13. Starting with a mixture of *cis*- and *trans*-2-hexene, outline a synthesis of *cis*-2-hexene.

14. 1-Pentene reacts with HBr to produce Product A which is treated with alcoholic KOH, resulting in Product B. Bromine in carbon tetrachloride is added to Product B, yielding Product C. This compound is then treated with sodium amide, resulting in the formation of Product D. Identify Products A, B, C, and D.

15. Starting with acetylene, ethyl bromide, and *n*-propyl bromide, outline a synthesis of (*E*)-3-heptene.

16. What product results when 3-hexyne undergoes hydroboration-oxidation, using BH_3 followed by alkaline H_2O_2?

17. Compound M is treated with strong base resulting in the formation of compound N. After reacting compound N with bromine in carbon tetrachloride, compound O results. O reacts with sodium amide in liquid ammonia and produces 1,4-pentadiyne. Identify compounds M, N, and O.

18. What is the name of the following dialkyne?

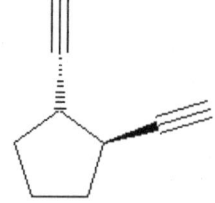

19. What product results when cyclodecyne undergoes ozonolysis?

20. What is the correct IUPAC name for the following alkyne?

$$CH_3-CH_2-\overset{\overset{\displaystyle CH_3}{|}}{CH}-C\equiv C-\overset{\overset{\displaystyle CH_3}{|}}{CH}-CH_3$$

   a. 2,3-dimethyl-3-heptyne

   b. 2,5-dimethyl-2-heptyne

   c. 2,5-dimethyl-3-heptyne

   d. 3,6-dimethyl-4-heptyne

21. Which of the following is the strongest base?

   a. $CH_3CH_2^-$

   b. $H_2C=CH^-$

   c. $HC\equiv C^-$

   d. NH_2^-

22. Which of the following sets of reagents could be used to convert 1-pentene to 1-pentyne?

   a. 1. Br_2 in CCl_4, 2. H_2SO_4/Hg^{2+}

   b. 1. Br_2 in CCl_4, 2. KOH in ethanol

   c. 1. Br_2 in CCl_4, 2. $NaNH_2$ in THF

   d. 1. HBr in CCl_4, 2. NaOH in ethanol

23. Which of the following could be used to convert a triple bond to a double bond?

   a. H_2 on a Pt catalyst

   b. 1. HBr, 2. $NaNH_2$

   c. H_2SO_4 and Hg^{2+}

   d. H_2 on Lindlar Pd

24. What product results when two moles of hydrogen chloride react with 4-methyl-2-pentyne?

   a. 2,2-dichloro-4-methylpentane

   b. 3,3-dichloro-4-methylpentane

   c. a and b

   d. none of these

25. Which product results when propyne reacts with aqueous sulfuric acid in the presence of mercury(II) cation catalyst?

   a. CH_3CH_2CHO

   b. CH_3COCH_3

   c. $CH_3CH(OH)CH_3$

   d. $CH_3CH_2CH_2OH$

26. What simple lab test could be performed to distinguish between 1-pentyne and 2-pentyne?

   a. The addition of Br_2 in CCl_4.

   b. The addition of H_2SO_4 and Hg^{2+}.

   c. The addition of Ag^+.

   d. The addition of H_2 on a Pt catalyst.

27. The reaction of an unknown alkyne with alkaline permanganate followed by the addition of acid produces the following two carboxylic acids.

What is the identity of the unknown alkyne?

   a. 2,6-dimethyl-4-octyne

   b. 2-methyl-4-octyne

   c. 2,4-dimethyl-4-octyne

   d. 2,6-dimethyl-3-octyne

28. Compound A results when sodium acetylide reacts with 1-bromobutane. One mole of Cl_2 reacts with one mole of A, producing compound B. What is the identity of B?

   a. *cis*-1,2-dichlorobutane

   b. *cis*-1,2-dichloro-1-hexene

   c. *trans*-1,2-dichloro-1-hexene

   d. *trans*-2,3-dichloro-2-hexene

29. Which of the following is the most stable alkyne?

    a. cyclopentyne

    b. cyclohexyne

    c. cycloheptyne

    d. cyclooctyne

30. What is the correct IUPAC name of the following?

    a. 2-propyn-1-ol

    b. 2-ethyn-1-ol

    c. prop-2-yne-1-ol

    d. none of these

31. An alkyne undergoes ozonolysis followed by hydrolysis and yields only 2-methylpropanoic acid, $CH_3CH(CH_3)COOH$. What is the identity of this alkyne?

    a. 2,4-dimethyl-2-hexyne

    b. 2,5-dimethyl-3-hexyne

    c. 2,4-dimethyl-3-hexyne

    d. none of these

32. What product results when one mole of 1-pentyne is first treated with one mole of HCl and then with one mole of HBr?

    a. 1-chloro-2-bromopentane

    b. 2-chloro-1-bromopentane

    c. 2-bromo-2-chloropentane

    d. 1-chloro-1-bromopentane

33. What product(s) can be isolated when 2-hexyne reacts with aqueous sulfuric acid and Hg^{2+}?

    a. 2-hexen-2-ol

    b. 2-hexen-3-ol

    c. 2-hexanone

    d. 3-hexanone

    e. a and b

    f. c and d

34. What is the product of the reaction of phenylacetylide, $C_6H_5-C\equiv C^-$, and *n*-butyl *p*-toluenesulfonate?

    a. 2-phenyl-1-hexyne

    b. 1-phenyl-1-hexene

    c. 1-phenyl-2-hexyne

    d. none of these

35. Which of the following results in the formation of 4,4-dimethyl-2-pentyne?

    a. $(CH_3)_3C^- Na^+ + HC\equiv CCH_3 \rightarrow$

    b. $(CH_3)_3C-I + CH_3C\equiv C^- Na^+ \rightarrow$

    c. $(CH_3)_3C-C\equiv C^- Na^+ + CH_3I \rightarrow$

    d. none of these

 # Check Yourself

1. The C–C triple bond is composed of one sigma and two pi bonds. This structure is linear. Each C atom is *sp* hybridized. (**C–C triple bond**)

2. As the bond order increases from single to triple the bond distance decreases. Hence, the C–C single bond is the longest and the C–C triple bond is shortest. The geometries associated with C–C single, double, and triple bonds are tetrahedral (109.5°), trigonal planar (120°), and linear (180°), respectively. (**Comparing C–C bonds**)

3. The hybridization of C atoms in C–C single bonds is sp^3, which means the *s* character is 25% (one *s* orbital out of a total of four). The hybridization of C atoms in C–C double bonds is sp^2, which means the *s* character is 33% (one *s* orbital out of a total of three). The hybridization of C atoms in C–C triple bonds is *sp*, which means the *s* character is 50% (one *s* orbital out of a total of two). (**C–C *s* character**)

4. The name of this alkyne is 2,2-dimethyl-3-hexyne. (**Alkyne nomenclature**)

5. The structure of 4-chloro-4-methyl-2-heptyne is as follows.

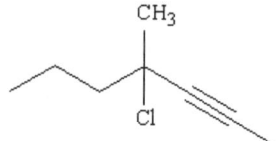

4-chloro-4-methyl-2-heptyne

(Alkyne nomenclature)

6. Acetylene, propyne, and 1-butyne are gases and the remaining lower-molecular-mass alkynes are liquids. They have similar trends in boiling points as alkanes and alkenes because they are nonpolar molecules. Alkynes have low densities and low water solubility. They are soluble in many nonpolar organic solvents. **(Properties of alkynes)**

7. The trend in acidity of acetylene, ethylene, and ethane is as follows.

    acetylene > ethylene > ethane

    Acids donate protons. When $HC\equiv CH$ donates a proton it produces the acetylide ion $HC\equiv C^-$. Because of the sp hybridization of the C atoms in the C–C triple bond (50% s character), the acetylide ion can more readily accept the negative charge. Ethylene and ethane only have 33% and 25% s character, and does not as readily accept the negative charge in the vinyl and ethyl anions. (**Acidity of acetylides**)

8. a. $NH_2^- + H-C\equiv C-H \rightleftharpoons NH_3 + H-C\equiv C^-$
    b. This equilibrium lies to the right because the amide ion is a stronger base than the acetylide ion and acetylene is a stronger acid than ammonia. (**Acidity of acetylides**)

9. $Na^+ HC\equiv C^- + CH_3CH_2-Br \rightarrow CH_3CH_2C\equiv CH$

    The acetylide ion is a nucleophile that will do an S_N2 displacement of a good leaving group such as bromide. (**Alkyne preparations**)

10. $Na^+ HC\equiv C^- + CH_3-Br \rightarrow CH_3C\equiv CH$
    $CH_3CCH + NaNH_2 \rightarrow Na^+ CH_3C\equiv C^-$
    $Na^+ CH_3C\equiv C^- + CH_3-Br \rightarrow CH_3C\equiv CCH_3$ (**Alkyne preparations**)

11. First dehydrate the alcohol, $(CH_3)_2CH-OH$, with sulfuric acid to produce propene, $CH_3CH=CH_2$. Next, add Br_2 in CCl_4 to brominate the alkene. This produces 1,2-dibromopropene, CH_3CBrCH_2Br. Finally, add $Na^+ NH_2^-$ in THF to fully dehydrohalogenate (double dehydrohalogenation) 1,2-dibromopropene to yield the product, propyne. (**Alkyne preparations**)

12. When alkynes are hydrated they initially produce enols, molecules with an OH group bonded to a double bond, which quickly tautomerize to ketones. Thus, 1-hexyne undergoes a hydration reaction, using aqueous sulfuric acid, H_2SO_4, and $HgSO_4$, and produces 2-hexanone. (**Alkyne reactions**)

$$CH_3-(CH_2)_3-C\equiv CH + H_2O \xrightarrow{H^+,\ Hg^{2+}} CH_3-(CH_2)_3-\overset{\overset{\displaystyle O}{\|}}{C}-CH_3$$

1-hexyne   2-hexanone

13. A good way to produce *cis*-2-hexene is to hydrogenate 2-hexyne using Lindlar's catalyst (Pd(Pb)). Thus, first halogenate the mixture of *cis*- and *trans*-2-hexene and then use $NaNH_2$ to produce 2-hexyne.
    (*cis* and *trans*) $CH_3CH=CHCH_2CH_2CH_3 + Br_2 \rightarrow CH_3CHBrCHBrCH_2CH_2CH_3$
    $CH_3CHBrCHBrCH_2CH_2CH_3 + NaNH_2 \rightarrow CH_3C\equiv CCH_2CH_2CH_3$
    $CH_3C\equiv CCH_2CH_2CH_3 + H_2$ (Lindlar's cat.) $\rightarrow$ *cis*-2-hexene **(Alkyne reactions)**

14. A = $CH_3CHBrCH_2CH_2CH_3$ **(Markovnikov addition)**
    B = $CH_3CH=CHCH_2CH_3$ **(Saytzeff elimination)**
    C = $CH_3CHBrCHBrCH_2CH_3$ **(Bromination)**
    D = $CH_3C\equiv CCH_2CH_3$ **(Alkyne reactions) (Alkyne synthesis)**

15. First, convert acetylene to sodium acetylide.

    $$NaNH_2 + HC\equiv CH \rightarrow Na^+\, HC\equiv C^-$$

    Next, react the sodium acetylide with *n*-propyl bromide.

    $$Na^+\, HC\equiv C^- + BrCH_2CH_2CH_3 \rightarrow HC\equiv CCH_2CH_2CH_3$$

    Then react the 1-pentyne with sodium amide to produce an acetylide.

    $$HC\equiv CCH_2CH_2CH_3 + Na^+\, NH_2 \rightarrow Na^+\, {}^-C\equiv CCH_2CH_2CH_3$$

    This product is then mixed with ethyl bromide to produce 3-heptyne.

    $$Na^+\, {}^-C\equiv CCH_2CH_2CH_3 + CH_3CH_2Br \rightarrow CH_3CH_2C\equiv CCH_2CH_2CH_3$$

    To produce the *E* or *trans* product, the 3-heptyne is treated with Na in liquid NH_3.

    $$CH_3CH_2C\equiv CCH_2CH_2CH_3 + Na(NH_3(l)) \rightarrow (E)\text{-3-heptene}$$

    (Alkyne synthesis)

16. Hydroboration-oxidation of alkynes initially produces the enol form of a carbonyl compound, which tautomerizes to the carbonyl compound. Thus, 1-hexyne produces 3-hexanone. **(Alkyne reactions)**

    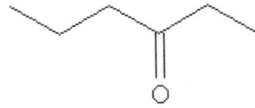

17. M = $BrCH_2CH_2CH_2CH_2CH_2Br$
    N = $H_2C=CHCH_2CH=CH_2$ **(Saytzeff elimination)**
    O = $BrCH_2CHBrCH_2CHBrCH_2Br$ **(Bromination) (Alkyne reactions)**

18. A $-C\equiv CH$ group is called an ethynyl group when attached to chains and rings. Thus, this compound is *trans*-1,2-diethynylcyclopentane. **(Alkyne nomenclature)**

19. Cyclodecyne is a ten-membered ring that contains a C–C triple bond. When cyclodecyne undergoes ozonolysis, the ring is cleaved and at the position of the double bond two carboxyl groups result. Hence, the product of the reaction is $HOOC(CH_2)_8COOH$. **(Alkyne reactions)**

20. c. 2,5-Dimethyl-3-heptyne is the correct answer because this molecule has seven C atoms with a triple bond located between the third and fourth C atoms. Additionally, two methyl groups are attached to the second and fifth C atoms. **(Alkyne nomenclature)**

21. a. The ethyl anion, $CH_3CH_2^-$, is the strongest base of the four selections because ethane, CH_3CH_3, is the weakest acid. Weaker acids have stronger conjugate bases than stronger acids. **(Alkyne acidity)**

22. c. First, add Br_2 in CCl_4 to brominate the double bond. Then add $NaNH_2$ in THF to double dehydrohalogenate the dihalide to 1-pentyne. (**Alkyne preparations**)

23. d. H_2 on Lindlar Pd hydrogenates the triple bond to a double bond and does not further reduce the double bond. (**Alkyne reactions**)

24. c. Both a and b are correct because 2,2-dichloro-4-methylpentane and 3,3-dichloro-2-methylpentane result from this reaction. (**Alkyne reactions**)

25. b. Acetone, CH_3COCH_3, results in this reaction because the enol $CH_3CH(OH)CH_3$ quickly tautomerizes to the keto form. (**Alkyne reactions**)

26. c. The addition of Ag^+ in ammonia produces an insoluble silver acetylide with 1-pentyne but not with 2-pentyne. (**Alkyne identification**)

27. a. During oxidative cleavage, the triple bond breaks and carboxyl groups are found on the C atoms that made up the C–C triple bond. Joining the molecules with a C–C triple bond at the carboxyl groups gives 2,6-dimethyl-4-octyne. (**Alkyne analysis**)

28. c. Sodium acetylide undergoes a nucleophilic substitution reaction with the 1-bromobutane to produce 1-hexyne, compound A. When one mole of Cl_2 adds across the double bond it produces 1,2-dichloro-1-hexene. Because halogens undergo anti addition across triple bonds, *trans*-1,2-dichloro-1-hexene results. (**Alkyne reactions**)

29. d. Cyclooctyne is the most stable molecule because C–C triple bonds cannot fit into smaller ring systems due to their linear geometry. The smallest unsubstituted ring that is stable at room conditions is cyclooctyne. (**Alkyne stability**)

30. a. 2-Propyn-1-ol is the correct name because this molecule has a chain of three C atoms. On the first C atom is an OH group and a triple bond is located between the second and third C atoms. (**Alkyne nomenclature**)

31. b. 2,5-Dimethyl-3-hexyne undergoes oxidative cleavage to produce 2-methylpropanoic acid. (**Alkyne analysis**)

32. c. 2-Bromo-2-chloropentane is the product because both hydrohalogenations follow Markovnikov's rule. (**Alkyne reactions**)

33. f. Both c and d are correct because the product of the hydration of an alkyne is a ketone. The carbonyl group can either form on the second or third C atom. (**Alkyne reactions**)

34. d. None of these are correct because the product is 1-phenyl-1-hexyne.

(**Alkyne reactions**)

35. c. The reaction of $(CH_3)_3C–C\equiv C^- \, Na^+$ and CH_3I results in the product. To use acetylides as nucleophiles they must attack methyl or primary halides. Acetylides are strong bases and initiate elimination reactions in secondary and tertiary halides. (**Alkyne reactions**)

Grade Yourself

Circle the number of questions you missed, then fill in the total incorrect for each topic. If you answered more than three questions incorrectly, you need to focus on that topic. (If a topic has less than three questions and you had at least one wrong, we suggest you study that topic also. Read your textbook, a review book, or ask your teacher for help.)

Subject: Alkynes

Topic	Question Numbers	Number Incorrect
C–C triple bond	1	
Comparing C–C bonds	2	
C–C *s* character	3	
Alkyne nomenclature	4, 5, 18, 20, 30	
Properties of alkynes	6	
Acidity of acetylides	7, 8	
Alkyne preparations	9, 10, 11, 22	
Alkyne reactions	12, 13, 16, 17, 19, 23, 24, 25, 28, 32, 33, 34, 35	
Alkyne synthesis	14, 15	
Alkyne acidity	21	
Alkyne identification	26	
Alkyne analysis	27, 31	
Alkyne stability	29	

Allylic and Conjugated Systems

9

Brief Yourself

The allylic cation, $H_2C=CHCH_2^+$, and allylic free radical, $H_2C=CHCH_2{}^\bullet$, are resonance stabilized by the delocalization of the π electrons in the double bond.

$$H_2C=CH-CH_2^+ \rightleftarrows H_2C^+-CH=CH_2$$

$$H_2C=CH-CH_2{}^\bullet+ \rightleftarrows {}^\bullet CH_2-CH=CH_2$$

As a result of resonance stabilization, allylic carbocations are more stable than tertiary carbocations, and allylic free radicals, are more stable than tertiary free radicals. As a result of the stability of allylic free radicals alkenes undergo allylic substitution reactions. A common way to accomplish this is to use *N*-bromosuccinimide, NBS, as a source of Br free radicals, Br$^\bullet$.

$$H_2C=CHCH_2 + NBS \rightarrow H_2C=CHCH_2-Br$$

The three classes of dienes are isolated, conjugated, and cumulated dienes (allenes). An isolated diene has double bonds that are separated by more than one sp^3 C atom. A conjugated diene system has two double bonds joined by a C–C single bond. A cumulated system has two double bonds to one C atom. 1,5-Hexadiene ($H_2C=CHCH_2CH_2CH=CH_2$), 1,3-hexadiene ($H_2C=CHCH=CHCH_2CH_3$), and 1,2-hexadiene ($H_2C=C=CHCH_2CH_2CH_3$), are examples of isolated, conjugated, and cumulated dienes, respectively.

Conjugated dienes are more stable than isolated dienes as a result of their π electron delocalization. A measure of this stability is called resonance or delocalization energy. The most stable conformations of dienes are the s-*cis* and s-*trans* conformations. The s-*cis* and s-*trans* conformations of 1,3-butadiene are as follows.

s-cis conformation   s-trans conformation

When dienes are protonated they react so as to produce an allylic cation. Because of the resonance of the allylic cation, dienes produce two products when they react with hydrogen halides. They yield both a 1,2-addition product and a 1,4-addition product. The 1,2-addition product results when the hydrogen halide undergoes a Markovnikov addition across one of the two double bonds—the direct addition. The 1,4-addition product results when the hydrogen bonds to C_1 and the halide bonds to C_4, the conjugate addition.

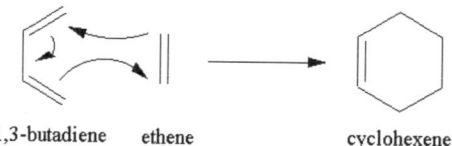

This reaction is under kinetic control at lower temperatures and is under thermodynamic control at higher temperatures. Halogens, X_2, (X = Cl and Br) undergo similar reactions with dienes producing 1,2- and 1,4-addition products.

Dienes (exclusively in the s-*cis* conformation) react with alkenes, alkynes, and other alkadienes (dienophiles), producing cyclohexenes and 1,4-cyclohexadiene derivatives. This is called the Diels-Alder reaction. The simplest Diels-Alder reaction is with 1,3-butadiene and ethene.

1,3-butadiene   ethene   cyclohexene

The Diels-Alder reaction is a concerted [4 + 2] cycloaddition reaction.

Test Yourself

1. a. What is a conjugated π electron system?

   b. Give an example of such a system.

   c. Draw the structure of this system.

2. Explain why the allylic cation is a conjugated π system.

3. a. Draw the structure of an allylic cation within a six-membered ring.

   b. Show the resonance in this cation.

4. A mixture of two alcohols results when 1-bromo-3-methyl-2-pentene is hydrolyzed with water and sodium carbonate. What alcohols are produced in this reaction? Explain what happens in this reaction.

5. Compare a conjugated diene to a cumulated diene. Give an example of each.

6. a. In what general class of dienes does the following molecule belong? Explain.

   b. What is the name of this diene?

7. Which of the following should have the lowest enthalpy of hydrogenation: 1,4-pentadiene or 1,3-pentadiene? Explain.

8. The C–C bond distance between the second and third C atoms in 1,3-butadiene is 146.3 pm. The C–C bond distance in ethane is 153.4 pm. Explain why the C2–C3 bond distance in butadiene is shorter than the bond distance in ethane.

(For problems 9 to 11, consider the s-*cis* and s-*trans* conformations of 1,3-butadiene.)

9. Draw the structures of the s-*cis* and s-*trans* conformations of 1,3-butadiene.

10. Explain how the s-*cis* and s-*trans* conformations of 1,3-butadiene differ.

11. Which of these structures is most stable? Explain.

(For problems 12 to 14, consider the reaction of HBr and 1,3-butadiene.)

12. What two products result when HBr (without the presence of organic peroxides) reacts with 1,3-butadiene?

13. Explain why two products result in the reaction of HBr and 1,3-butadiene.

14. At low temperatures, the 1,2-addition product of the reaction of HBr and 1,3-butadiene is the major product of the reaction. At higher temperatures, the 1,4-addition product is the major product. Write a complete explanation.

15. Draw the structures of the products that result when Cl_2 reacts with 1,3-cyclohexadiene.

16. a. In what class of reaction does the Diels-Alder reaction belong?

    b. What names are given to the two reactants in a Diels-Alder reaction?

17. Write the mechanism for the Diels-Alder reaction using the reactants 1,3-butadiene and ethene.

18. Draw the structure of the product of the reaction of 1,3-butadiene and *cis*-2-butene.

19. What Diels-Alder reaction will produce the following Diels-Alder adduct?

20. Which of the following carbocations is most stable?

    a. $CH_3CH=CHCH_2^+$

    b. $CH_3CH_2CH_2CH_2^+$

    c. $CH_3CH=C^+CH_3$

    d. $CH_3CH_2CH_2^+$

21. What is the name of the following diene?

    a. (*E, E*)-2,4-hexadiene

    b. (*E, Z*)-2,4-hexadiene

    c. (*Z, Z*)-2,4-hexadiene

    d. none of these

22. Which of the following has the lowest enthalpy of hydrogenation?

    a. 1,2-octadiene

    b. 2,3-octadiene

    c. 2,4-octadiene

    d. 2,5-octadiene

23. Which of the following is a chiral molecule?

    a. 2,3-octadiene

    b. 2,4-octadiene

    c. 2,4-dimethyl-1,3-hexadiene

    d. none of these

24. If one mole of HBr first reacts with 1,3-butadiene followed by one mole of Cl_2, what is the product(s) of the reaction?

    a. 3-bromo-1,2-dichlorobutane

    b. 2-bromo-1,2-dichlorobutane

    c. 2-bromo-1,3-dichlorobutane

    d. 1-bromo-2,3-dichlorobutane

    e. b and c

    f. a and d

25. What products result when one mole of Cl_2 reacts with 1,3-cyclohexadiene?

    a. 1,2-dichlorocyclohexane

    b. 1,3-dichlorocyclohexane

    c. 3,4-dichlorocyclohexane

    d. 3,6-dichlorocycohexane

    e. a and b

    f. c and d

    g. none of these

26. The enthalpy of hydrogenation for 1-pentene is +126 kJ/mol. If the enthalpy of hydrogenation for 1,3-pentadiene is +230 kJ/mol, estimate the resonance (delocalization) energy of 1,3-pentadiene.

    a. 252 kJ

    b. 104 kJ

    c. 22 kJ

    d. cannot be calculated from this information

27. Consider the following structure.

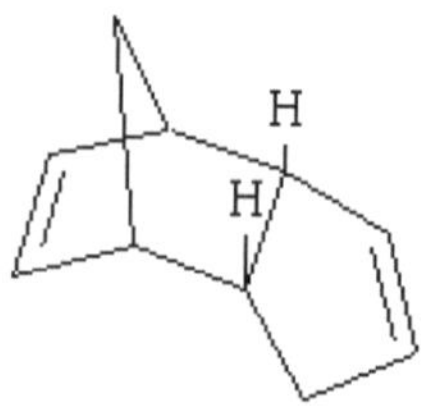

    Which of the following sets of compounds undergoes a Diels-Alder reaction to produce this structure?

    a. 1,3-cyclopentadiene and 1,3-cyclopentadiene

    b. 1,3-cyclopentadiene and 1-methylcyclobutene

    c. 1,3-cyclohexadiene and cyclobutene

    d. none of these

28. What is the product of the reaction of one mole of NBS (*N*-bromosuccinimide) and cyclohexene in carbon tetrachloride with light?

    a. 1-bromocyclohexane

    b. 1-bromocyclohexene

    c. 3-bromocyclohexene

    d. none of these

29. What is the name of the following compound?

    $H_2C=C=CH_2$

    a. propyne

    b. allene

    c. 1,3-propanediene

    d. allodiene

30. Which molecular orbitals are filled in 1,3-butadiene?

    a π_1 only

    b. π_1 and π_2

    c. π_1, π_2, and π_3

    d. π_1, π_2, π_3, and π_4

31. Which of the following is **not** a product of the reaction of one mole of Br_2 with 1,3,5-hexatriene?

    a. 5,6-dibromo-1,3-hexadiene

    b. 3,6-dibromo-1,4-hexadiene

    c. 1,6-dibromo-2,4-hexadiene

    d. 2,6-dibromo-1,3-hexadiene

32. Compound Y undergoes a Diels-Alder reaction with itself. When this Diels-Alder adduct is catalytically hydrogenated, it produces ethylcyclohexane. What is the identity of compound Y?

    a. 1,3-butadiene

    b. 1,3-pentadiene

    c. 2-methyl-1,3-butadiene

    d. none of these

33. Which of the following is **not** correct about the Diels-Alder reaction?

    a. It is a stereospecific reaction.

    b. It is a concerted [4 + 2] cycloaddition reaction.

    c. It requires an s-*cis* diene and dienophile.

    d. The products are either cyclohexane or 1,4 cyclohexadiene derivatives.

34. Which of the following is **not** the product of the reaction of one mole of HI and 2-methyl-1,3-butadiene?

    a. 1-iodo-3-methyl-2-butene

    b. 3-iodo-3-methyl-1-butene

    c. 3-iodo-2-methyl-1-butene

    d. 1-iodo-2-methyl-2-butene

    e. 3-iodo-2-methyl-2-butene

35. Compound X has the formula of C_7H_{10}. Upon catalytic hydrogenation one mole of X adds two moles of H_2. When X undergoes ozonolysis it produces the following compounds.

    What is the identity of compound X?

    a. 5-methyl-1,3-hexadiene

    b. 5-methyl-1,3-cyclohexadiene

    c. 3-methyl-1,3-cyclohexadiene

    d. none of these

✔ Check Yourself

1. a. A conjugated π electron system is one in which π electrons are delocalized over four or more atoms. An isolated double bond is not a conjugated system because the π electrons are localized between two C atoms.
   b. An example of such a system is a conjugated diene, two alternating double bonds.
   c. The general structure for conjugated dienes is as follows. (**Conjugate systems**)

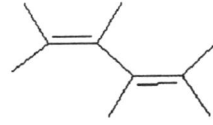

Conjugated Diene

2. The allylic cation has the structure $H_2C=CHCH_2^+$. This structure has a continuous π system because the C atom that carries the charge has an empty p orbital. Thus, the structure of the allylic cation is as follows. (**Allylic cations**)

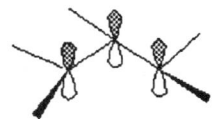

3. a. The structure of this allylic cation is as follows.

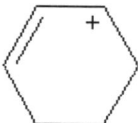

b. The resonance in this cation is as follows.

(Allylic cations)

4. $CH_3CH_2C(CH_3)=CHCH_2Br \rightarrow CH_3CH_2C(CH_3)=CHCH_2OH + CH_3CH_2COH(CH_3)CH=CH_2$
    The products of this reaction are 3-methyl-2-penten-1-ol and 3-methyl-1-penten-3-ol. They result because when the leaving group, Br, departs, the following allylic cation is produced.

$$CH_3CH_2C(CH_3)=CHCH_2^+ \leftrightarrow CH_3CH_2C^+(CH_3)CH=CH_2$$

Thus, the OH group can bond to either the first or third C atom. **(Allylic reactions)**

5. A conjugated diene is one that has two alternating double bonds. An example of a conjugated diene is 1,3-butadiene. A cumulated diene is one that has two double bonds to one C atom. An example of a cumulated diene is 1,2-butadiene. **(Diene structure)**

6. a. This is an isolated diene because the double bonds are separated by more than one sp^3 C atom.
    b. The name of this molecule is 1,4-cycloheptadiene. **(Diene structure)**

7. 1,3-Pentadiene has a lower enthalpy of hydrogenation than 1,4-pentadiene. 1,3-Pentadiene is a conjugated diene system and 1,4-pentadiene is an isolated diene system. Conjugated diene systems are more stable as a result of their resonance energy, also called delocalization energy. **(Diene stability)**

8. The structure of 1,3-butadiene, $H_2C=CH–CH=CH_2$, shows that this single bond is between two sp^2 hybridized C atoms. The single bond in ethane is between two sp^3 C atoms. The s character of sp^2 C atoms is higher than that of sp^3 C atoms and thus they attract electrons more strongly; thus, the bond distance between the second and third C atoms in 1,3-butadiene is shorter than the one in ethane. **(Diene structure)**

9. The s-*cis* and s-*trans* conformations of 1,3-butadiene are as follows.

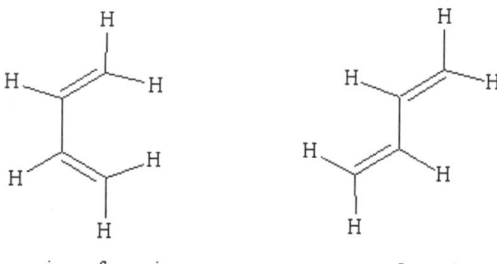

s-cis conformation   s-trans conformation

(Diene structure)

10. The s-*cis* conformation differs from the s-*trans* conformation with respect to the rotation around the C_2–C_3 bond. Because the π system must be coplanar, the first and fourth C atoms of 1,3-butadiene can either be on the same side, s-*cis*, or on opposite sides, s-*trans*. **(Diene structure)**

11. The s-*cis* conformation is less stable than the s-*trans* conformation because of the van der Waals repulsions of the H atoms bonded to the first and fourth C atoms. **(Diene structure)**

12. The two products that form are 3-bromo-1-butene, the 1,2-addition product, and 1-bromo-2-butene, 1,4-addition product. **(Diene reactions)**

13. 3-Bromo-1-butene and 1-bromo-2-butene result because the positive charge is delocalized across the conjugated π system. The bromide ions can bond to either C atom that carries the positive charge. **(Diene reactions)**

14. The 1,2-addition product, 3-bromo-1-butene, is under kinetic control, and the 1,4-addition product, 1-bromo-2-butene, is under thermodynamic control. This means that the activation energy to produce the 1,2-addition product is less than that of the 1,4-addition product, and that the 1,4-addition product is lower in energy than the 1,2-addition product. Hence, at lower temperatures the pathway to the kinetically controlled product, 1,2-addition product, will be the major product, but at higher temperatures, the 1,4-addition product will be produced in greatest amount. **(Diene reactions)**

15. The products of the reaction are as follows. **(Diene reactions)**

3,4-dichlorocyclohexene   3,6-dichlorocyclohexene

16. a. The Diels-Alder reaction is classified as a concerted [4 +2] cycloaddition reaction.
    b. The two reactants in the Diels-Alder reaction are the diene and the molecule that attacks the diene, the dienophile. Dienophiles are usually substituted or unsubstituted alkenes and alkynes. **(Diels-Alder reaction)**

17. The Diels-Alder reaction mechanism is concerted. Ethene adds to 1,3-butadiene and produces cyclohexene. **(Diels-Alder reaction)**

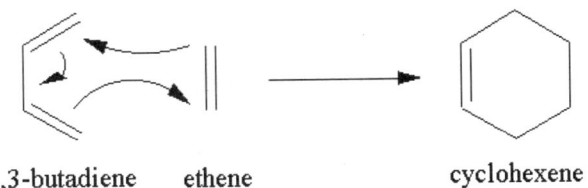

1,3-butadiene   ethene   cyclohexene

18. The product of the reaction is *cis*-4,5-dimethylcyclohexene. **(Diels-Alder reaction)**

19.

1,3-cyclopentadiene   2-butyne

(Diels-Alder reaction)

20. a. $CH_3CH=CHCH_2^+$ is the most stable cation because it is an allylic cation. **(Allylic cations)**

21. b. (*E, Z*)-2,4-Hexadiene is the name for this molecule because one double bond is the *E* isomer (*trans*) and the other is the *Z* isomer (*cis*). **(Diene structures)**

22. c. 2,4-Octadiene has the lowest enthalpy of hydrogenation because it is a conjugated diene. All of the others are either cumulated or isolated dienes, which are less stable and release more heat when hydrogenated. **(Diene stability)**

23. a. 2,3-Octadiene is an allene that is not superimposable on its mirror image; thus, it is a chiral molecule. **(Diene structure)**

24. f. Both a (3-bromo-1,2-dichlorobutane) and d (1-bromo-2,3-dichlorobutane) are correct. In this reaction, HBr can undergo both 1,2 addition and 1,4 addition. Then the Cl_2 adds across the remaining double bond. **(Diene reactions)**

25. g. None of these are correct because the products that result from this reaction are 3,4-dichlorocyclohexene and 3,6-dichlorocyclohexene. **(Diene reactions)**

26. c. The resonance energy for 1,3-pentadiene can be estimated by comparing its enthalpy of hydrogenation, –230 kJ/mol, with that of two double bonds that do not interact. This is obtained by doubling the value for 1-pentene, 2 mol × –126 kJ/mol, or –252 kJ. Thus, 1,3-pentadiene releases 22 kJ less (–230 kJ – (–252 kJ)) energy than two isolated double bonds. This is an estimate of the resonance energy. **(Diene stability)**

27. a. The compound is the dimer of 1,3-cyclopentadiene. **(Diels-Alder reaction)**

28. c. 3-Bromocyclohexene is the product because NBS is a source of bromine free radicals, which undergo an allylic substitution. **(Allylic reactions)**

29. b. Allene is the common name for this molecule. 1,2-Propadiene is its IUPAC name. **(Allenes)**

30. b. 1,3-Butadiene has a total of four π electrons. These electrons fill the two lowest energy molecular orbitals of 1,3-butadiene, π_1 and π_2. **(Diene structure)**

31. d. 2,6-Dibromo-1,3-hexadiene is not a product of this reaction. 5,6-dibromo-1,3-hexadiene is the 1,2-addition product, 3,6-dibromo-1,4-hexadiene is the 1,4-addition product, and 1,6-dibromo-2,4-hexadiene is the 1,6-addition product. **(Triene reactions)**

32. a. 1,3-Butadiene undergoes a Diels-Alder reaction and produces 4-vinylcyclohexene, which is catalytically hydrogenated to ethylcyclohexane. **(Diels-Alder reaction)**

33. d. The products are either cyclohexene or 1,4-cyclohexadiene derivatives. **(Diels-Alder reaction)**

34. e. 3-Iodo-2-methyl-2-butene is not formed in this reaction. Selections a and b result from the carbocation $(CH_3)_2C^+CH=CH_2$, and selections c and d result from $H_2C=C(CH_3)C^+HCH_3$. **(Diene reactions)**

35. b. Knowing that the molecule adds two moles of H_2 indicates that the molecule is a diene. Having another degree of unsaturation shows that it must be a cyclic compound. Placing the two ozonolysis products gives the placement of the double bonds and the attached methyl group. Compound X is 5-methyl-1,3-cyclohexadiene. **(Diene identification)**

Grade Yourself

Circle the number of questions you missed, then fill in the total incorrect for each topic. If you answered more than three questions incorrectly, you need to focus on that topic. (If a topic has less than three questions and you had at least one wrong, we suggest you study that topic also. Read your textbook, a review book, or ask your teacher for help.)

Subject: Allylic and Conjugated Systems

Topic	Question Numbers	Number Incorrect
Conjugate systems	1	
Allylic cations	2, 3, 20	
Allylic reactions	4, 28	
Diene structure	5, 6, 8, 9, 10, 11, 12, 13, 14, 15, 21, 23, 30	
Diene stability	7, 22, 26	
Diene reactions	24, 25, 34	
Diene identification	35	
Diels-Alder reaction	16, 17, 18, 19, 27, 32, 33	
Allenes	29	
Triene reactions	31	

Aromatic Compounds and Aromaticity

10

Brief Yourself

Arenes are aromatic compounds that exhibit similar properties to benzene, C_6H_6, and its derivatives. The benzene molecule can be represented by the following two resonance structures.

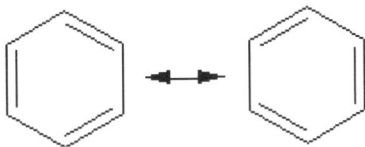

Benzene is a conjugated triene that has a high resonance energy, which makes it more stable. Each C atom in benzene is sp^2 hybridized and all C–C bond distances are equal. Benzene principally undergoes substitution reactions and not addition reactions like alkenes.

The stability of benzene can also be explained using the molecular orbital theory. Benzene has six molecular orbitals, three bonding and three antibonding orbitals, which result from the overlap of its six $2p$ orbitals. The three bonding orbitals are π_1, π_2, and π_3, and the three antibonding orbitals are π_4, π_5, and π_6. The three bonding orbitals are lower in energy than the electrons in isolated double bonds, and the three antibonding orbitals are higher in energy. Benzene has six electrons in its π system, thus, according to the aufbau principle, these electrons will occupy the lower energy bonding orbitals. The difference in energy between the six electrons in the bonding orbitals compared with the same electrons in six $2p$ C orbitals equals the resonance energy for benzene.

Hückel's rule is used to predict if a compound will exhibit aromatic properties. Hückel's rule states that monocyclic fully conjugated polyenes are aromatic if they are planar and have $4n + 2$ π electrons, where n is a positive integer such as 0, 1, 2, 3 . . . Hückel's rule can also be extended to polycyclic systems. Benzene is a molecule that has six π electrons, in which n equals 1. An example of an aromatic molecule that has 10 π electrons is naphthalene, a polycyclic aromatic.

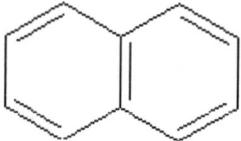

naphthalene

Ions can also be aromatic since they follow Hückel's rule. The cyclopropenyl cation is an aromatic ion with two π electrons ($n = 0$).

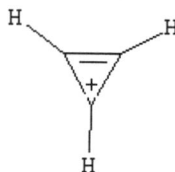

cyclopropenyl cation

Heterocyclic compounds can also be aromatic. Pyrrole is an aromatic heterocyclic amine.

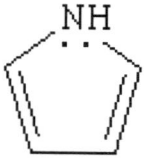

pyrrole

The N atom in pyrrole is sp^2 hybridized and the lone pair is part of the conjugated π system.

Some monosubstituted benzene derivatives are named by adding the substituent name to the word "benzene." Examples of such molecules are nitrobenzene ($C_6H_5–NO_2$), chlorobenzene ($C_6H_5–Cl$), and ethylbenzene ($C_6H_5–CH_2CH_3$). Some monosubstituted benzene derivatives have their own unique names. Examples of such molecules include toluene (methylbenzene), styrene (vinylbenzene), phenol (hydroxybenzene, $C_6H_5–OH$), and benzoic acid ($C_6H_5–COOH$).

Disubstituted benzenes often use the prefixes *ortho (o)*, *meta (m)*, and *para (p)*. *Ortho* means that the substituent groups are in the first and second positions. *Meta* means that the substituent groups are in the first and third positions, and *para* means that the substituent groups are in the first and fourth positions. The following shows the three isomers of dibromobenzene.

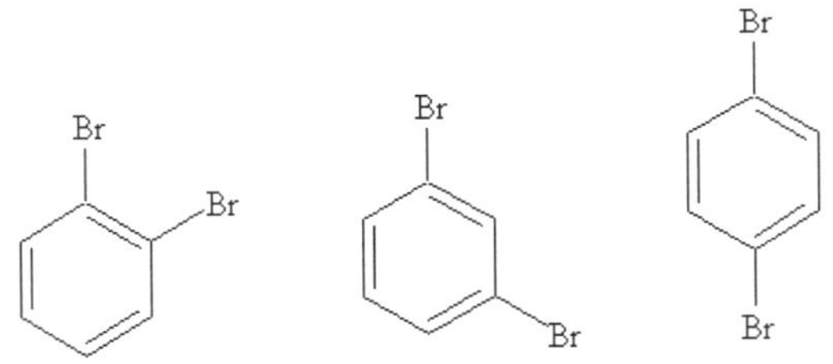

Br
1,2-dibromobenzene
o-dibromobenzene

Br
1,3-dibromobenzene
m-dibromobenzene

Br
Br
1,4-dibromobenzene
p-dibromobenzene

Removing a H from benzene produces a phenyl group, C_6H_5-. The phenyl group can be a substituent on a chain or ring. 1,1-Diphenylcyclohexane is an example of such a molecule.

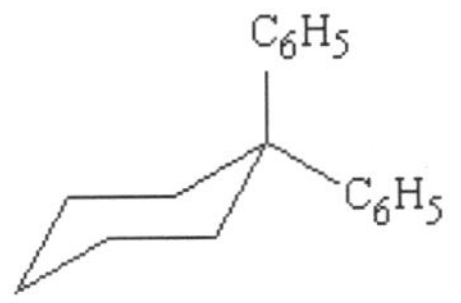

C_6H_5
C_6H_5

Electron delocalization stabilizes both the benzylic cation and free radical.

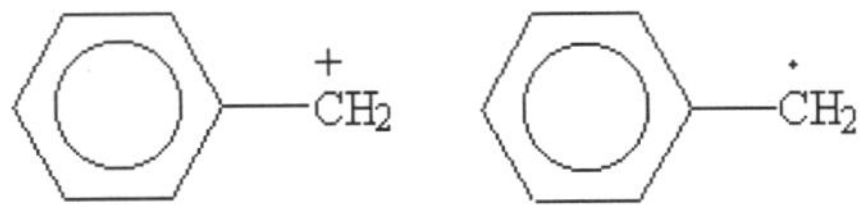

$\overset{+}{C}H_2$

$\overset{\cdot}{C}H_2$

benzylic cation   benzylic free radical

In the presence of light, both Cl_2 and Br_2 undergo a free-radical substitution at the benzylic position of alkyl substituted benzene compounds. Free-radical bromination can also take place using NBS with benzoyl peroxide.

Alkyl substituted benzene derivatives can be oxidized by permanganate or chromic acid to benzoic acid, C_6H_5-COOH. For example, *n*-propyl benzene, $C_6H_5-CH_2CH_2CH_3$ is oxidized to benzoic acid.

Test Yourself

1. a. Draw two resonance structures of benzene.

    b. How is the structure of benzene often written?

2. Explain how the resonance energy of benzene can be calculated.

3. Explain the stability of benzene using the molecular orbital theory.

4. a. State Hückel's rule.

    b. How can it be applied to benzene?

    c. What are the "magic numbers" of π electrons in aromatic systems?

5. a. Draw the structure of *cis, trans, cis, cis, trans*-[10]-annulene.

    b. Use Hückel's rule to decide if [10]-annulene is aromatic.

    c. Explain why [10]-annulene is unstable and does not exhibit the properties of aromatic compounds.

(For problems 6 to 8, consider the molecule *cis, trans, trans, cis, trans, trans, cis, trans, trans*-[18]-annulene. [18]-Annulene has been synthesized and the inside of the molecule is large enough so that the interior H atoms do not repel significantly.)

6. Draw the structure of [18]-annulene.

7. Use Hückel's rule to decide if [18]-annulene is aromatic.

8. Considering that C–C double bonds are shorter than C–C single bonds, describe the bond distances between C atoms in [18]-annulene.

9. Cyclopentadiene is almost as strong an acid as water. Explain the acidity of cyclopentadiene in terms of the stability of its conjugate base.

10. Draw the structure of the cyclopropenyl cation and determine if it is aromatic. Explain.

11. The molecule furan is composed of a heterocyclic ring that contains one O atom.

    a. Draw the structure of furan.

    b. Explain why furan is aromatic.

12. Draw the structures of two polycyclic aromatics that have three six-member rings.

13. a. Write the names of all constitutional isomers of xylene, dimethylbenzene.

    b. Describe the similarities and differences in physical properties of these isomers.

14. Write the name for the following compound.

15. Draw the structure of *p*-nitrobiphenyl.

16. Draw four resonance structures that show the electron delocalization of the benzyl carbocation.

17. What is the product of the free radical bromination of ethylbenzene using NBS with benzoyl peroxide in CCl_4?

18. What product results from the oxidation of *p*-ethyltoluene using chromic acid and heat?

19. a. Draw the structure of the cyclobutadienyl dication.

    b. Draw the π molecular orbital diagram for this ion and determine if it is aromatic.

20. How many π molecular orbitals are completely filled in the ground state of benzene?

    a. none
    b. one
    c. three
    d. six

21. Which of the following is true about the cycloheptatrienyl free radical?

    a. It has $4n + 2$ π electrons.
    b. It is an aromatic free radical.
    c. It is an isolatable stable free radical.
    d. none of these

22. Which of the following reagents could be used to convert cyclooctatetraene to the cyclooctatetraene dianion?

    a. $KMnO_4$
    b. K
    c. KBr
    d. none of these

23. Consider the following molecule.

    Which of the following is a correct statement regarding this molecule?

    a. This molecule is an aromatic because it has 6 π electrons.
    b. This molecule will become an aromatic ion if it loses the Br atom and retains the pair of electrons that bond it to the ring.
    c. This molecule will become an aromatic ion if it loses the Br atom and the pair of electrons that bond it to the ring.
    d. none of these

24. Which of the following best describes the structure of 1,3-cyclobutadiene?

    a. aromatic

    b. antiaromatic

    c. nonaromatic

    d. none of these

25. Which of the following molecules is a polycyclic aromatic hydrocarbon?

    a. naphthalene

    b. pyridine

    c. furan

    d. b and c

26. Consider the structures of pyrrole and pyridine.

    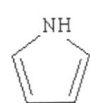 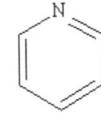

    pyrrole   pyridine

    Which of the following is a correct statement about pyrrole and pyridine?

    a. Both pyrrole and pyridine are strongly basic because both have N atoms that can readily accept protons.

    b. Both pyrrole and pyridine are very weak bases because both can weakly accept an electron pair.

    c. Pyrrole is a weaker base than pyridine because when it accepts a proton it loses its aromatic properties.

    d. Pyridine is a weaker base than pyrrole because the lone pair electrons on its N atom are part of the aromatic electron system.

27. What is the name of the following compound?

    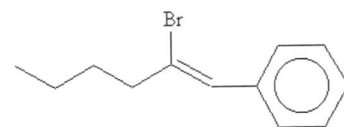

    a. (Z)-2-bromo-1-phenyl-1-hexene

    b. 2-bromohexylbenzene

    c. 1-phenyl-2-bromohexane

    d. none of these

28. *n*-Butylbenzene reacts with Br₂ and light and produces Compound A. A reacts with KOH in alcohol and produces Compound B. B reacts with HBr in the presence of organic peroxides to produce Compound C. What is the identity of compound C?

    a. 1-bromo-1-phenylbutane

    b. 3-bromo-1-phenylbutane

    c. 2-bromo-1-phenylbutane

    d. none of these

29. Which of the following is a correct name for the following compound?

    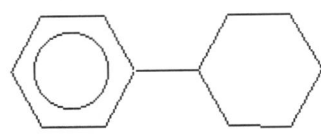

    a. hexylbenzene

    b. biphenyl

    c. cyclohexyl benzene

    d. phenyl benzene

    e. dicyclohexane

30. What is the product of the reaction of styrene, vinyl benzene, when it is completely hydrogenated using H₂ and a Pt catalyst?

    a. ethylcyclohexane

    b. ethyl benzene

    c. vinyl cyclohexene

    d. none of these

31. Which of the following best describes the molecule pentalene?

    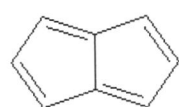

    a. aromatic

    b. antiaromatic

    c. nonaromatic

    d. none of these

32. Which of the following has the highest normal boiling point?

    a. fluorobenzene

    b. chlorobenzene

    c. bromobenzene

    d. iodobenzene

33. Which of the following sets of reagents will convert 1-bromo-2-phenylethane to 1-phenylethanol?

    a. 1. $NaOCH_2CH_3$/CH_3CH_2OH, 2. BH_3/diglyme, 3. H_2O_2, OH^-

    b. H_2SO_4 and heat

    c. 1. $NaNH_2$/NH_3 2. Br_2/light, 3. H^+/H_2O

    d. 1. $NaOC(CH_3)_3$/$(CH_3)_3COH$, 2. $Hg(OAc)_2$/THF-H_2O, 3. $NaBH_4$, OH^-

34. If one mole of H_2 is added to the following compound, which double bond will be hydrogenated?

    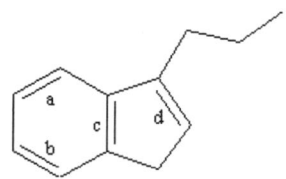

    a. a

    b. b

    c. c

    d. d

    e. none of these

35. Consider the following compounds, cycloheptatrienone and cyclopentadienone.

    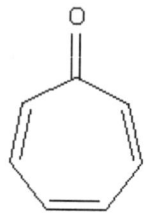   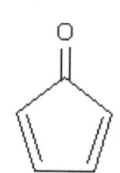

    cycloheptatrienone   cyclopentadienone

    Which of the following correctly describes these compounds?

    a. Cycloheptatrienone is a stable compound and cyclopentadienone is unstable.

    b. Cyclopentadienone is a stable compound and cycloheptatrienone is unstable.

    c. Both of these compounds are stable compounds.

    d. Neither cyclopentadienone nor cycloheptatrienone is stable.

 **Check Yourself**

1. a. Two resonance structures of benzene are as follows.

   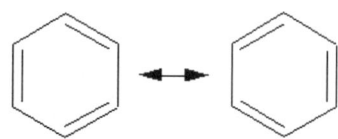

   b. To show the resonance in benzene, a circle is drawn inside of a six-membered ring. **(Benzene structure)**

   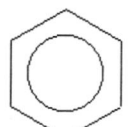

2. One way to estimate the resonance or delocalization energy for benzene is to compare the enthalpy of hydrogenation of benzene with that of the hypothetical molecule 1,3,5-cyclohexatriene or the real molecule (*Z*)-1,3,5-hexatriene. The amount of energy liberated by benzene is significantly smaller than either of these molecules. The difference in energy is the resonance energy. (**Benzene stability**)

3. There are six molecular orbitals, three bonding and three antibonding orbitals, which result from the six $2p$ orbitals in the π system of benzene. The three bonding orbitals are π_1, π_2, and π_3, and the three antibonding orbitals are π_4, π_5, and π_6. The three bonding orbitals are lower in energy than the $2p$ orbitals and the three antibonding are higher in energy. Benzene has six electrons in its π system, thus, according to the aufbau principle they will occupy the lower energy bonding orbitals. The difference in energy between the six electrons in the bonding orbitals compared to the same electrons in double bonds that do not have resonance is the resonance energy for benzene. (**Benzene stability**)

4. a. Hückel's rule states that monocyclic fully conjugated polyenes are aromatic if they are planar and have $4n + 2$ π electrons, where n is a positive integer.
   b. The benzene ring is planar and monocyclic and has 6 π electrons in a conjugated system.
   c. When n equals 0, 1, 2, and 3, the "magic numbers" of π electrons in aromatic systems are 2, 6, 10, and 14, respectively. (**Hückel's rule**)

5. a. The structure of [10]-annulene is as follows.

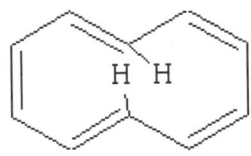

[10]-annulene

(cis, trans, cis, cis, trans)

   b. It has 10 π electrons and would be expected to be aromatic if it has a planar structure.
   c. While this compound has 10 π electrons the repulsions of the two interior H atoms force the structure out of a planar configuration which is required for aromatic compounds. (**Aromatic properties**)

6. The structure of [18]-annulene is as follows.

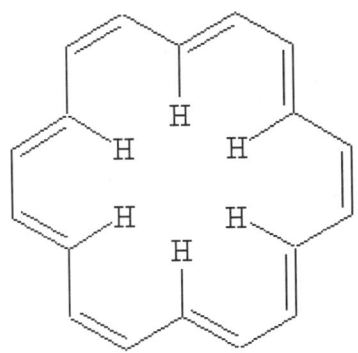

(**Annulene properties**)

7. [18]-Annulene follows Hückel's rule because it is planar and has $4n + 2$ π electrons (n is 4); hence, it is aromatic. (**Aromatic properties**)

8. Because [18]-annulene is aromatic, the bond distances for all C–C bonds are between a single and a double C–C bond. (**Aromatic properties**)

9. The following equilibrium shows the acid-ionization of cyclopentadiene, producing a proton and the cyclopentadienide anion.

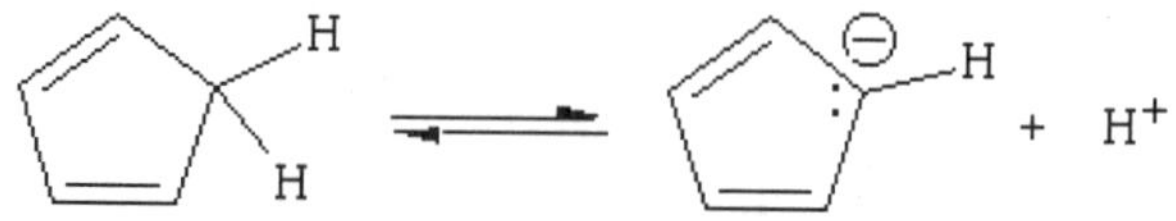

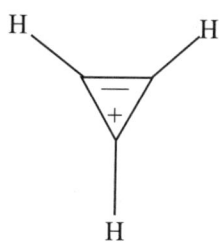

cyclopentadienide anion

The cyclopentadienide anion is aromatic because it has 6 π electrons in a planar ring. The resonance stabilization as a result of being an aromatic anion shifts the equilibrium to the right, making cyclopentadiene significantly more acidic than most dienes. (**Aromatic ions**)

10. The structure of the cyclopropenyl cation is as follows. It has 2 π electrons, which makes it an aromatic cation ($n = 0$). (**Aromatic ions**)

11. a. The structure of furan is as follows.

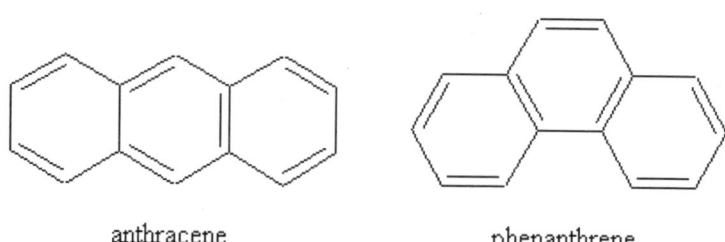

furan

b. It is a heterocyclic aromatic compound because one of the lone pairs on the O atom contributes to the π system, giving a total of 6 π electrons. (**Heterocyclic aromatics**)

12. Two example of polycyclic aromatic compounds are anthracene and phenanthrene. (**Polycyclic aromatics**)

anthracene   phenanthrene

13. a. The three isomers of xylene are *o*-xylene, 1,2-dimethylbenzene, *m*-xylene, 1,3-dimethylbenzene, and *p*-xylene, 1,4-dimethylbenzene.
b. Because these are constitutional isomers, they have similar but different physical properties. For example, they have different boiling points and refractive indices. (**Benzene derivatives**)

14. The name of this compound is 2,6-dichloro-4-nitrophenol. (**Aromatic nomenclature**)

15. The structure of *p*-nitrobiphenyl is as follows. (**Aromatic nomenclature**)

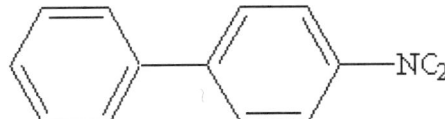

16. Four resonance structures for the benzyl carbocation are expressed as follows.

Benzyl Carbocation

17. Substitution occurs at the benzylic position when alkyl substituted benzenes react with bromine free radicals. Thus, the product of the reaction is 1-bromo-1-phenylethane. (**Benzylic substitution**)

18. Alkyl substituted benzenes are oxidized to carboxylic acids. Thus, the disubstituted 4-ethyltoluene is oxidized to *p*-benzenedicarboxylic acid. (**Benzylic reactions**)

19. a. The structure of the cyclobutadienyl dication is as follows.

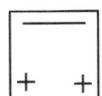

b. The MO diagram for this cation is as follows.

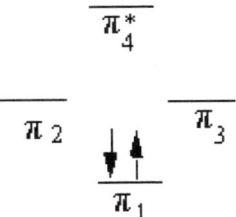

The two electrons fill the lowest energy π bonding orbital; thus, this dication is aromatic. Hückel's rule would also predict that this ion is aromatic because it has 2 π electrons in a planar ring. (**MOs and aromatics**)

20. c. The three lowest energy bonding π molecular orbitals, π_1, π_2, and π_3, are filled. The three higher energy antibonding molecular orbitals are not filled. (**MOs and aromatics**)

21. d. None of these is correct because free radicals have an odd number of electrons and cannot obey Hückel's rule. (**Aromatic properties**)

22. b. A strong reducing agent such as the alkali metal K is used to add two electrons to cyclooctatetraene to produce the aromatic cyclooctatetrene dianion. (**Aromatic ions**)

23. c. This molecule becomes an aromatic ion if it loses the Br atom and the pair of electrons that bond it to the ring. It becomes the cycloheptatrienyl cation which has a continuous π cloud with six π electrons. (**Aromatic ions**)

24. b. 1,3-Cyclobutadiene is classified as antiaromatic because it has a planar ring with conjugated double bonds, which is less stable then the corresponding acyclic unsaturated species. (**Antiaromatic species**)

25. a. Naphthalene is the polycyclic aromatic compound with the formula $C_{10}H_8$ with two fused six-member rings. (**Polycyclic aromatics**)

26. c. Pyrrole is a weaker base than pyridine because when pyrrole accepts a proton it loses its aromatic properties. When it accepts a proton, the sp^2 N atom is converted to a sp^3 N, which makes the molecule nonaromatic. (**Aromatic properties**)

27. a. (Z)-2-Bromo-1-phenyl-1-hexene is the name of this compound because it has a Br atom on the second C atom and a phenyl group on the first C atom of a 1-hexene chain. (**Aromatic nomenclature**)

28. c. Compound C is 2-bromo-1-phenylbutane. When *n*-butylbenzene reacts with bromine free radicals, the Br free radical substitutes at the benzylic position, producing 1-bromo-1-phenylbutane (A). Treatment with KOH in alcohol produces 1-phenyl-1-butene (B). HBr does an anti-Markovnikov addition across the double bond in the presence of organic peroxides. (**Substituted benzene reactions**)

29. c. This molecule has a cyclohexyl group bonded to a benzene ring. Thus, it is cyclohexyl benzene. (**Aromatic nomenclature**)

30. a. Catalytic hydrogenation of styrene adds four moles of H_2 to the molecule, giving ethylcyclohexane. (**Aromatic reactions**)

31. b. Pentalene is antiaromatic because it has eight π electrons and the ring is planar. (**Antiaromatic species**)

32. d. Iodobenzene has the highest boiling point because it has the greatest number of polarizable electrons, which produces the strongest London forces. (**Aromatic properties**)

33. d. In the first reaction, $NaOC(CH_3)_3/(CH_3)_3COH^-$ converts the bromide to styrene through an E2 elimination. Next, $Hg(OAc)_2/THF-H_2O$ followed by $NaBH_4/OH^-$ produces the desired product through Markovnikov addition of water across the double bond. (**Substituted benzene reactions**)

34. d. The double bond that most readily hydrogenates is the one outside of the aromatic ring. If more H_2 were available, all of the double bonds would be hydrogenated. (**Aromatic reactions**)

35. a. Cycloheptatrienone is a stable compound and cyclopentadienone is so unstable that it has never been isolated. Cycloheptatrienone has a resonance structure that forms the aromatic cycloheptatrienyl ion. Cyclopentadienone has four π electrons and is antiaromatic. (**Aromatic properties**)

Grade Yourself

Circle the number of questions you missed, then fill in the total incorrect for each topic. If you answered more than three questions incorrectly, you need to focus on that topic. (If a topic has less than three questions and you had at least one wrong, we suggest you study that topic also. Read your textbook, a review book, or ask your teacher for help.)

Subject: Aromatic Compounds and Aromaticity

Topic	Question Numbers	Number Incorrect
Benzine structure	1	
Benzine stability	2, 3	
Hückel's rule	4	
Aromatic properties	5, 7, 8, 21, 26, 32, 35	
Annulene properties	6	
Aromatic ions	9, 10, 22, 23	
Heterocyclic aromatics	11	
Polycyclic aromatics	12, 25	
Benzene derivatives	13	
Aromatic nomenclature	14, 15, 27, 29	
Benzylic substitution	17	
Benzylic reactions	18	
MOs and aromatics	19, 20	
Antiaromatic species	24, 31	
Substituted benzene reactions	28, 33	
Aromatic reactions	30, 34	

Electrophilic Aromatic Substitution

11

 **Brief Yourself**

Benzene and its derivatives undergo electrophilic aromatic substitution reactions. The general equation for this reaction is as follows

$$C_6H_5-H + E-Y \rightarrow C_6H_5-E + HY$$

in which E–Y is a reagent that contains the electrophile E.

When an electrophile attacks a benzene ring, it produces a carbocation. The rate-determining step for electrophilic aromatic substitution reactions is the first step, the formation of the carbocation, a cyclohexadienyl cation.

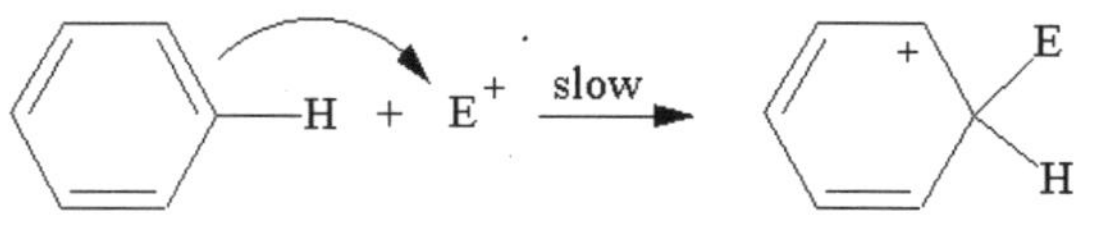

cyclohexadienyl cation

In the second step, the base Y⁻ bonds to the H⁺ lost as the aromatic ring is regenerated.

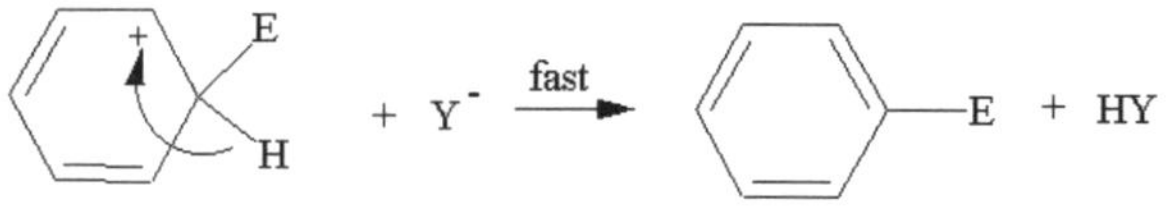

The principal types of electrophilic aromatic substitution reactions are nitration (substitutes $-NO_2$), halogenation (substitutes –X), sulfonation (substitutes $-SO_3H$), alkylation (substitutes –R), and acylation (substitutes –COR). These reactions only differ in the electrophile that attacks the ring and the reagents needed to generate them.

In the nitration reaction, the nitronium ion, NO_2^+, is the electrophile that bonds to the ring. It forms in a mixture of nitric acid and sulfuric acid.

In the halogenation reaction, either Cl_2 or Br_2, X_2, with a Lewis acid such as FeX_3 is used.

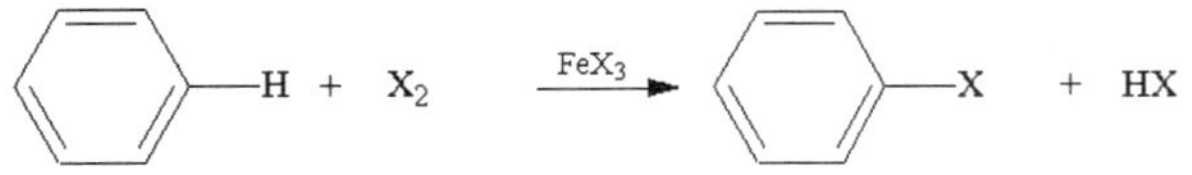

In this reaction, the Lewis acid weakens the bond between the halogen atoms and produces an electrophilic halogen atom that attacks the ring.

In the sulfonation reaction, fuming sulfuric acid, a mixture of SO_3 and H_2SO_4, is used to produce benzenesulfonic acid.

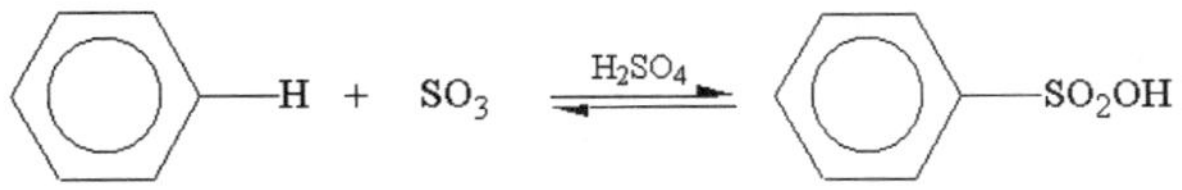

The electrophile in the sulfonation reaction is the SO_3 molecule. This reaction differs from the other electrophilic aromatic substitution reactions because it is reversible.

In the Friedel-Crafts alkylation reaction, an alkyl halide and Lewis acid are used to generate a carbocation that attacks the ring, producing an alkyl-substituted benzene.

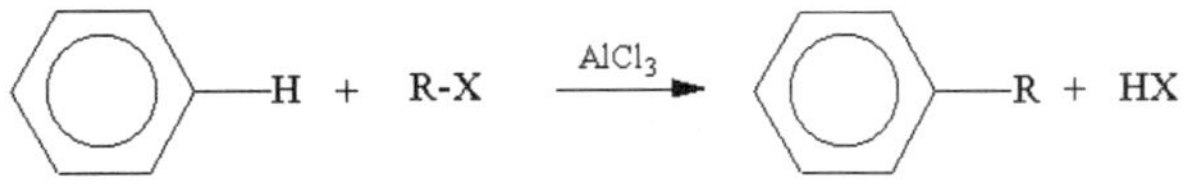

In this reaction, the alkyl halide must be secondary or tertiary because primary halides may rearrange to more stable secondary and tertiary carbocations.

In the Friedel-Crafts acylation reaction, an acyl chloride or acid anhydride and a Lewis acid produce aromatic ketones, acylbenzenes. The electrophile that attacks the ring is the resonance-stabilized acylium ion, $R-C\equiv O^+$.

$$\text{C}_6\text{H}_5\text{-H} + \text{R}-\overset{\text{O}}{\underset{\text{Cl}}{\text{C}}} \xrightarrow{\text{AlCl}_3} \text{C}_6\text{H}_5-\overset{\text{O}}{\text{C}}-\text{R} + \text{HCl}$$

The carbonyl group produced in this reaction can be reduced to a methylene group, CH_2, by either the Clemmensen reduction, using a zinc-mercury amalgam and HCl, or the Wolff-Kishner reduction, using hydrazine and strong base in a solvent of triethyleneglycol.

Electrophilic aromatic substitution of monosubstituted benzene derivatives depends on the substituent group. Some groups activate the ring (activators), making it more reactive, and others deactivate the ring (deactivators), making it less reactive. Some common aromatic ring activators are $-NH_2$, $-NHR$, $-NR_2$, $-OH$, $-OR$, $-NHCOR$, $-OCOR$, $-R$, and $-Ar$. Some common deactivators are $-NO_2$, $-CF_3$, $-SO_3H$, $-CN$, $-COCl$, $-COOR$, $-COOH$, $-COR(H)$, and $-X$. Groups that activate the aromatic ring direct the attacking electrophile to the *ortho* and *para* positions, *ortho-para* directors. Groups that deactivate the aromatic ring direct the attacking electrophile to the *meta* position, *meta* directors. One exception to this is are halides, $-X$. While they are weakly deactivating, they are *ortho-para* directors. This may be explained in terms of their ability to donate electron density into the ring through their lone pair electrons.

Test Yourself

1. a. What general type of reactions do benzene and benzene derivatives undergo?

   b. Write a general equation for this reaction.

2. a. What type of high-energy intermediate is produced in electrophilic aromatic substitution reactions?

   b. Write the equation for the rate-determining step for electrophilic aromatic substitution reactions. Explain what happens in this step.

3. a. Write the equation for the nitration of benzene.

   b. What is the actual electrophile that attacks the benzene ring? How is it generated?

4. a. What is the product of the reaction of sulfuric acid and benzene? Draw its structure.

   b. How is this reaction different from the nitration of benzene?

   c. What electrophile attacks the benzene ring?

5. a. Why is a Lewis acid, such as $FeBr_3$, required to catalyze the bromination reaction of benzene?

   b. What is the product of the bromination of benzene?

6. a. What product(s) results when $(CH_3)_3C-Cl$ reacts with benzene in the presence of a Lewis acid catalyst?

   b. What is the name given to this reaction?

7. Explain why the product of the reaction of isobutyl chloride and benzene with an $AlCl_3$ catalyst is *t*-butylbenzene and not isobutylbenzene. Write an equation that supports your answer.

8. What product results when cyclopentene and benzene react in H_2SO_4?

9. Write the equation for the Friedel-Crafts acylation of benzene with CH_3COCl.

10. Write equations that show a synthesis that produces *n*-hexylbenzene in good yield.

11. What is the principal product(s) that results when ethylbenzene undergoes nitration with nitric acid in sulfuric acid? Explain.

12. What is the principal product that results when nitrobenzene undergoes chlorination with Cl_2 and $AlCl_3$? Explain.

13. Compare the products that result when benzene first is nitrated and then brominated, to first brominating benzene and then nitrating the product.

14. Compound *A* results when benzene reacts with benzoyl chloride, C_6H_5COCl, with a Lewis acid catalyst. Compound *B* results when *A* reacts with a nitric acid-sulfuric acid mixture. Compound *C* results when *B* reacts with Zn(Hg) in HCl. Identify the structures of *A*, *B*, and *C*.

15. a. Draw three resonance structures that result from the *ortho* attack on toluene by bromine with a $FeBr_3$ catalyst.

    b. Which of these structures is most stable?

16. How could deuterobenzene, C_6H_5–D, be synthesized? D is the symbol for deuterium, ^2H.

17. Draw the structure of the product of the bromination of the following compound. Explain.

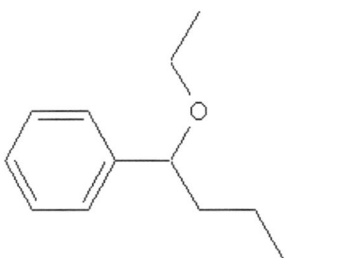

18. Compound *W* results when benzene reacts with $(CH_3)_3CCH_2COCl$ and $AlCl_3$. *W* reacts with a mixture of nitric acid and sulfuric acid to produce Compound *X*. Treating *X* with Zn(Hg) and HCl produces compound *Y*. Compound *Z* results when *Y* is heated in concentrated $KMnO_4$ followed by acid hydrolysis. Draw the structures of *W*, *X*, *Y*, and *Z*.

19. Starting with benzene show how the following compound, 1-ethoxy-1-phenylpentane, can be synthesized.

20. What electrophile attacks the benzene ring during a nitration reaction?

    a. HNO₃

    b. $H_2NO_3^+$

    c. NO₂

    d. NO_2^-

    e. NO_2^+

21. Why is an aluminum chloride catalyst used in Friedel-Crafts alkylation reactions?

    a. It lowers the energy of the benzene ring, making it more reactive.

    b. It bonds with one of the H atoms on the benzene ring, weakening the bond.

    c. It accepts a pair of electrons from the alkyl halide, making it easier to form a carbocation.

    d. It reacts with the halogen in the alkyl halide, stabilizing the C–X bond.

22. Which of the following reagents cannot be used to generate electrophiles in the Friedel-Crafts alkylation reaction?

    a. alkene and HF

    b. alkyl halide and AlCl₃

    c. alcohol and H₂SO₄

    d. alcohol and BF₃

    e. none of the above

23. Which of the following is **not** true about the trifluoromethyl group, CF_3-, as a substituent on an aromatic ring?

    a. The CF_3- group deactivates the ring.

    b. The CF_3- group activates the ring.

    c. The CF_3- is an *ortho-para* director.

    d. The CF_3- is a *meta* director.

    e. a and b

    f. b and c

24. Consider the following molecule.

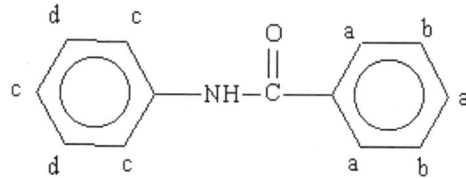

    At what position will sulfonation principally occur when this molecule is treated with SO_3 and H_2SO_4?

25. Which of the following is **not** a limitation of Friedel-Crafts alkylation reactions?

    a. Rearrangements can occur with structures that will not give the desired product.

    b. The addition of strongly activating groups makes multiple alkylations difficult to avoid.

    c. They can only be accomplished with benzene and its derivatives with activators. Strongly deactivating groups do not react under these conditions.

    d. all of these

26. What is the major product of the reaction of 2-chloropropane with excess nitrobenzene, using an $AlCl_3$ catalyst?

    a. *m*-isopropylnitrobenzene

    b. *o*-isopropylnitrobenzene

    c. *p*-isopropylnitrobenzene

    d. none of these

27. Consider the structure of naphthalene, a polycyclic aromatic hydrocarbon.

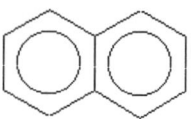

naphthalene

    How many different products result when naphthalene undergoes a bromination reaction with Br_2 and $FeBr_3$?

    a. one

    b. two

    c. three

    d. no reaction occurs

28. What is the principal product of the reaction of *m*-bromotoluene when it is treated with acetyl chloride and $AlCl_3$?

    a. 4-bromo-2-methylacetophenone

    b. 2-bromo-4-methylacetophenone

    c. 5-bromo-3-methylacetophenone

    d. 6-bromo-1-methylacetophenone

    e. a and b

    f. c and d

29. What is the product of the reaction of nitrobenzene and CH_3Cl with an $AlCl_3$ catalyst?

    a. a mixture of *o*-nitrotoluene and *p*-nitrotoluene

    b. *m*-nitrotoluene only

    c. *p*-nitrotoluene only

    d. no reaction takes place

30. Which of the following is the most strongly activating group for electrophilic aromatic substitution?

    a. $-OCH_3$

    b. $-CF_3$

    c. $-Cl$

    d. $-C_6H_5$

31. Which of the following is the most strongly deactivating group for electrophilic aromatic substitution?

    a. $-OH$

    b. $-NH_2$

c. –COCl

d. –Br

32. What is the principal product(s) when chlorobenzene reacts with Br_2 and $FeBr_3$?

   a. *o*-bromochlorobenzene only

   b. *m*-bromochlorobenzene only

   c. *p*-bromochlorobenzene only

   d. both *o*-bromochlorobenzene and *p*-bromo-chlorobenzene

33. Which of the following would produce a reasonable yield of *p*-nitrobenzoic acid?

   a. Nitration of benzoic acid

   b. Nitration of toluene, separate the *p*-isomer, and oxidize the methyl group with chromic acid.

   c. Friedel-Crafts acylation of toluene, separate the *p*-isomer, and reduce the carbonyl with the Wolff-Kishner reduction

   d. none of these

34. What single isomer would be found in greatest yield when isopropyl chloride undergoes a Friedel-Crafts alkylation with *t*-butylbenzene?

   a. *o*-isopropyl-*t*-butylbenzene

   b. *m*-isopropyl-*t*-butylbenzene

   c. *p*-isopropyl-*t*-butylbenzene

   d. none of these

35. Which of the following would be the major product of the reaction of *m*-nitrotoluene with ethyl chloride, using an aluminum chloride catalyst?

   a. 2-ethyl-3-nitrotoluene

   b. 4-ethyl-3-nitrotoluene

   c. 5-ethyl-3-nitrotoluene

   d. none of these

✔ Check Yourself

1. a. Benzene and its derivatives undergo electrophilic aromatic substitution reactions.
    b. The general equation for this reaction is as follows

$$C_6H_5\text{–H} + \text{E–Y} \rightarrow C_6H_5\text{–E} + \text{HY}$$

in which E–Y is a reagent that contains the electrophile E. (**Electrophilic aromatic substitution**)

2. a. When an electrophile attacks a benzene ring it produces a carbocation.
    b. The rate determining step is the first step, the formation of the carbocation.

cyclohexadienyl cation

The carbocation produced is the resonance-stabilized cyclohexadienyl cation. (**Electrophilic aromatic substitution**)

3. a. The nitration reaction for benzene is as follows.

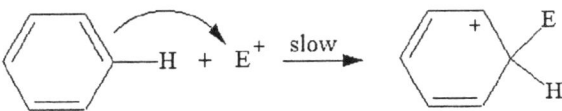

b. The electrophile that attacks the ring is the nitronium cation, NO_2^+, which is produced from the reaction of nitric acid and sulfuric acid. (**Nitration**)

4. a. The product of the reaction is benzenesulfonic acid.

benzenesulfonic acid

   b. The sulfonation reaction of benzene is a reversible reaction. The nitration of benzene is not.
   c. The electrophile that attacks the benzene ring is SO_3. (**Sulfonation**)

5. a. A Lewis acid catalyst such as $FeBr_3$ forms a Lewis acid-base adduct that makes one of the Br atoms more electrophilic.

   $Br–Br + FeBr_3 \rightarrow Br–Br^\delta{}^+ \!\!-Fe^\delta{}^- \!\!-Br_3$

   In this adduct, the Fe carries a negative formal charge and the adjacent Br carries a positive formal charge.
   b. The product of the reaction is bromobenzene, $C_6H_5–Br$. (**Halogenation**)

6. a. The product of this reaction is t-butylbenzene, $C_6H_5–C(CH_3)_3$.
   b. This reaction is called the Friedel-Crafts alkylation reaction. (**Friedel-Crafts alkylation**)

7. This is a Friedel-Crafts alkylation reaction. In the first step of the mechanism the $AlCl_3$ combines with the isobutyl chloride. The adduct of the isobutyl chloride and $AlCl_3$ rearranges through a hydride shift and produces the more stable t-butyl cation, which is the electrophile that attacks the benzene ring. (**Friedel-Crafts alkylation**)

8. The product of the reaction is cyclopentyl benzene, C_6H_5-C_5H_9, because the sulfuric acid protonates the cyclopentene and produces the cyclopentyl cation which undergoes a Friedel-Crafts alkylation reaction with the benzene. (**Friedel-Crafts alkylation**)

9. The equation for the Friedel-Crafts acylation reaction of benzene and acetyl chloride is as follows.

(**Friedel-Crafts acylation**)

10. To accomplish this synthesis the Friedel-Crafts acylation reaction must be used followed by reduction of the resulting carbonyl group using the Clemmensen reduction. A Friedel-Crafts alkylation cannot be used because rearrangements would occur in a primary halide. First, perform a Friedel-Crafts acylation of benzene using $CH_3CH_2CH_2CH_2CH_2COCl$ and $AlCl_3$.

   $C_6H_5–H + CH_3CH_2CH_2CH_2CH_2COCl \rightarrow C_6H_5–COCH_2CH_2CH_2CH_2CH_3$

   Then use $Zn(Hg)$ and HCl to reduce the carbonyl group to a CH_2 group.

   $C_6H_5–COCH_2CH_2CH_2CH_2CH_3 + Zn(Hg)/HCl \rightarrow C_6H_5–CH_2CH_2CH_2CH_2CH_2CH_3$ (**Friedel-Crafts acylation**)

11. An ethyl group, as all R groups, is an *ortho-para* director; hence, the principal products of this reaction are *o*- and *p*-nitroethyl benzene. (**Orientation of electrophilic aromatic substitution**)

12. The nitro group is a *meta* director; thus, the principal product of this reaction is *m*-chloronitrobenzene. (**Orientation of electrophilic aromatic substitution**)

13. When benzene is first brominated it produces bromobenzene. The nitration of bromobenzene produces both *o*- and *p*-bromonitrobenzene. Reversing the order first gives nitrobenzene, which has a *meta*-directing nitro group; hence, the principal product is *m*-bromonitrobenzene. (**Orientation of electrophilic aromatic substitution**)

14. Benzene reacts with benzoyl chloride in the presence of $AlCl_3$ to produce benzophenone (*A*).

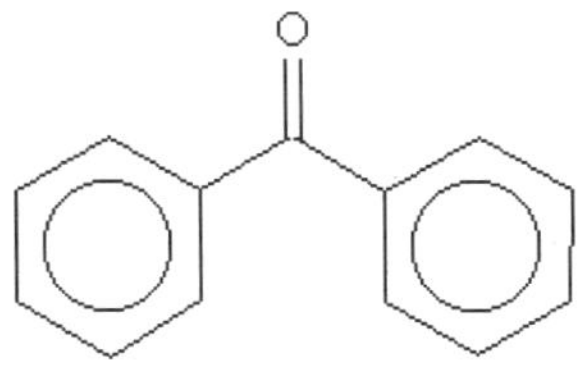

Benzophenone undergoes a nitration reaction to produce *m*-nitrobenzophenone.

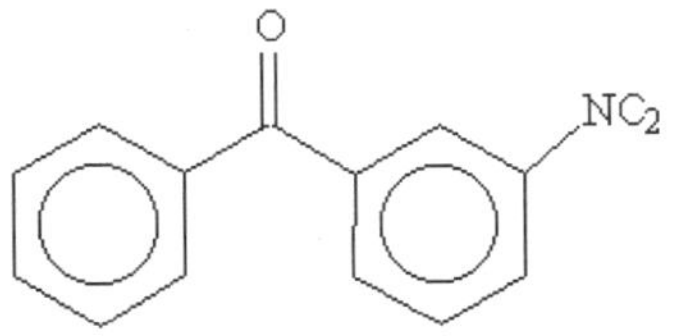

m-Nitrobenzophenone is reduced to phenyl *m*-nitrophenyl methane.

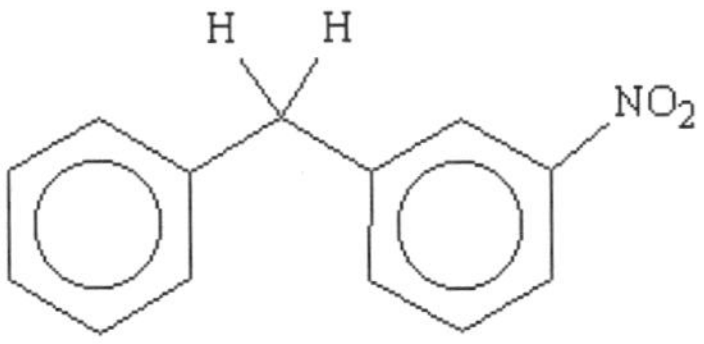

(**Electrophilic aromatic substitution reactions**)

15. a. The three resonance structures are as follows.

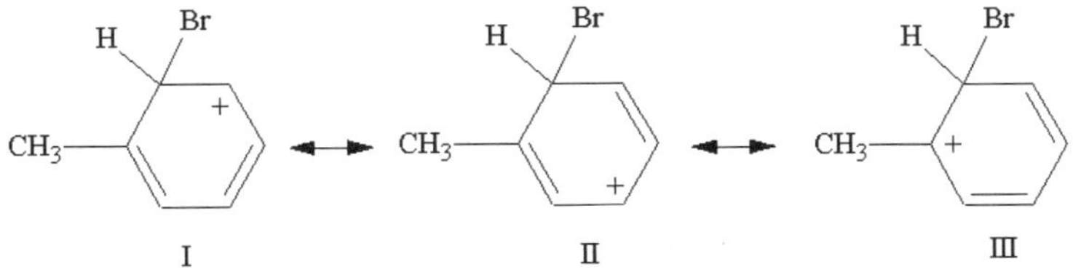

   b. Structure III is the lowest-energy resonance structure because the electron donating methyl group decreases the positive charge on the adjacent C atom. (**Orientation of electrophilic aromatic substitution**)

16. Deuterobenzene can be synthesized by first sulfonating benzene with SO_3 in H_2SO_4, and then removing the -SO_3H group using D_2O with catalytic amounts of acid, D^+. (**Sulfonation**)

   $C_6H_5-SO_3H + D_2O \ (D^+) \rightarrow C_6H_5-D$

17. The NO_2 group deactivates its ring; thus, the Br will attack the other ring. Because the CH_3O- group activates the ring, the principal product will be the *ortho*-substituted product. (**Orientation of electrophilic aromatic substitution**)

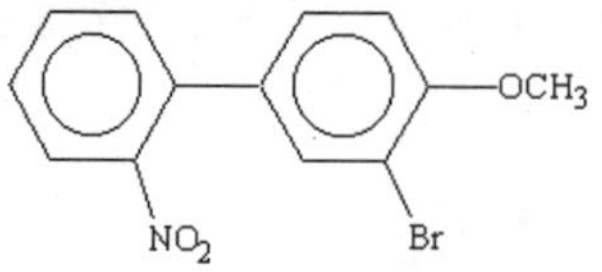

18.   *W* is produced in a Friedel-Crafts acylation reaction.

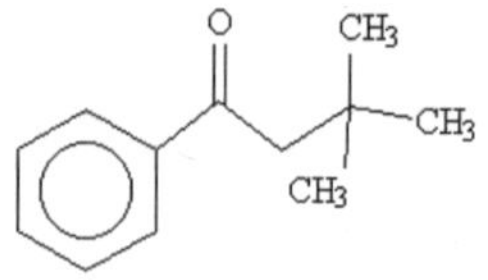

X results when the ring is nitrated at the *meta* position.

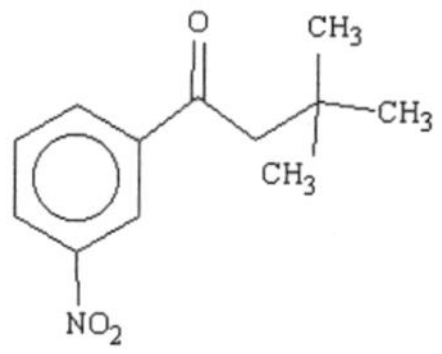

Y results when the carbonyl group is reduced to a methylene group.

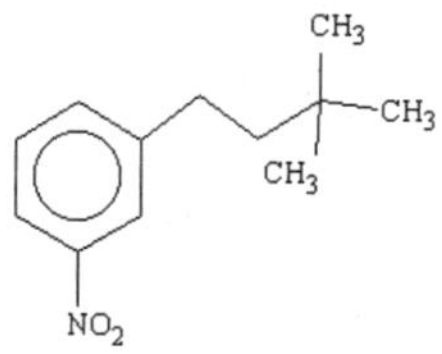

Z results when the alkyl side chain is oxidized to a carboxylic acid.

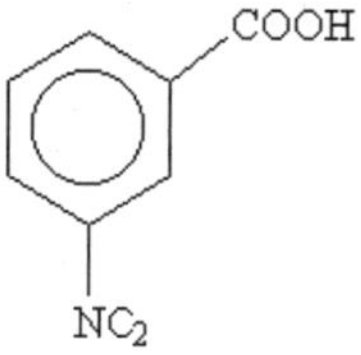

(Electrophilic aromatic substitution reactions)

19. The synthesis can be accomplished as follows.

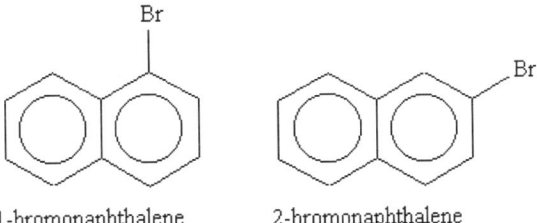

(Aromatic synthesis)

20. e. The nitronium ion, NO_2^+, is the electrophile that bonds to the ring in nitration reactions. **(Nitration)**

21. c. It accepts a pair of electrons from the alkyl halide, making it easier to form a carbocation. For secondary and tertiary alkyl halides, the $AlCl_3$ abstracts the Cl^-, forming $AlCl_4^-$ and a carbocation. **(Friedel-Crafts alkylation)**

22. e. None of the above is the correct answer because all of the sets of reagents can be used in Friedel-Crafts alkylation reactions. **(Friedel-Crafts alkylation)**

23. f. The trifluoromethyl group readily pulls electrons from the benzene ring which makes it a deactivator and a *meta* director. **(Substituent effects)**

24. c. The ring bonded to the –NH– group is activated; thus, sulfonation occurs at the *ortho-para* positions. The ring bonded to the –CO– group is deactivated and sulfonation does not occur to any great extent. **(Orientation of electrophilic aromatic substitution)**

25. d. All of the listed limitations are problems encountered with Friedel-Crafts alkylation reactions. **(Friedel-Crafts alkylation)**

26. d. The nitro group strongly deactivates the ring not allowing Friedel-Crafts reactions to occur. **(Friedel-Crafts alkylation)**

27. b. Two products result, 1-bromonaphthalene and 2-bromonaphthalene. The principal product is 1-bromonaphthalene because it forms a more stable carbocation. The structures of 1-bromonaphthalene and 2-bromonaphthalene are as follows. **(Halogenation)**

28. e. Both 4-bromo-2-methylacetophenone and 2-bromo-4-methylacetophenone form because both substituents direct *ortho* and *para* to each other.

(Orientation of electrophilic aromatic substitution)

29. d. No reaction takes place because the strongly deactivating nitro group does not allow the Friedel-Crafts alkylation reaction to occur. **(Friedel-Crafts alkylation)**

30. a. The methoxy group, –OCH$_3$, is the strongest activating group because it can donate a pair of electrons into the ring through resonance. **(Ring activation/deactivation)**

31. c. The acyl chloride group, –COCl, is the strongest deactivating group because it can withdraw electron density from the ring through resonance. **(Ring activation/deactivation)**

32. d. Both *o*-bromochlorobenzene and *p*-bromochlorobenzene result because the Cl group is an *ortho-para* director. **(Orientation of electrophilic aromatic substitution)**

33. b. Because both groups, nitro and carboxyl group, are *meta* directors, an *o-p* director must first be added to the ring. The nitration of toluene produces both *o*- and *p*-nitrotoluene; hence, the *p*-isomer must be separated from the *o*-isomer. Finally, the methyl group is oxidized to a carboxyl group using chromic acid. **(Aromatic synthesis)**

34. c. The *t*-butyl group is an *ortho-para* director. Nonetheless, *p*-isopropyl-*t*-butylbenzene is found in greatest yield because the *t*-butyl group blocks the attack of the isopropyl group at the ortho position. **(Orientation of electrophilic aromatic substitution)**

35. b. 4-Ethyl-3-nitrotoluene is the principal product. The methyl group in toluene is an *ortho-para* director and the nitro group is a *meta* director. Usually, when both an *ortho-para* and *meta* director are attached to a ring, the principal product is *para* to the *ortho-para* director and *ortho* to the *meta* director. However, steric affects will diminish the attack of the C between the two groups. **(Orientation of electrophilic aromatic substitution)**

Grade Yourself

Circle the number of questions you missed, then fill in the total incorrect for each topic. If you answered more than three questions incorrectly, you need to focus on that topic. (If a topic has less than three questions and you had at least one wrong, we suggest you study that topic also. Read your textbook, a review book, or ask your teacher for help.)

Subject: Electrophilic Aromatic Substitution

Topic	Question Numbers	Number Incorrect
Electrophilic aromatic substitution	1, 2	
Nitration	3, 20	
Sulfonation	4, 16	
Halogenation	5, 27	
Friedel-Crafts alkylation	6, 7, 8, 21, 22, 25, 26, 29	
Friedel-Crafts alcylation	9, 10	
Orientation of electrophilic aromatic substitution	11, 12, 13, 15, 17, 24, 28, 32, 34, 35	
Electrophilic aromatic substitution reactions	14, 18	
Aromatic synthesis	19, 33	
Substituent effects	23	
Ring activation/deactivation	30, 31	

IR Spectroscopy, NMR Spectroscopy, and Mass Spectrometry

12

Brief Yourself

The IR region of the electromagnetic spectrum used to analyze organic molecules extends from 2.5×10^{-6} m to 1.7×10^{-5} m. Spectroscopists often measure this region of the IR spectrum using wavenumbers, the reciprocal of the wavelengths. The IR region extends from 4000 to 600 cm^{-1}.

Infrared radiation excites molecules from their ground vibrational energy state to excited vibrational energy states. To produce an IR absorption, a vibration of a bond must change its dipole moment. This is called IR active. Vibrations that do not change the dipole moment are IR inactive. The C–C double bond stretch in 1-propene is an IR active vibration. The C–C double bond stretch in ethene is an IR inactive vibration.

Bonds vibrate in two ways: stretching and bending. Each bond stretches and bends at specific frequencies in the IR range. Whenever this IR energy is absorbed, a peak is found in the spectrum. In the range of 2700 to 3500 cm^{-1} the C–H, O–H, and N–H stretches are found. From 2100 to 2300 cm^{-1}, both the C–C and C–N triple bonds absorb. From 1600 to 1800 cm^{-1}, the C–C, C–O, and C–N double bonds absorb. Finally, in the range of 650 to 800 cm^{-1}, aromatic absorptions are found.

[1]H nuclear magnetic resonance spectroscopy results from the quantized nuclear spin states of H nuclei (protons). If a strong magnetic field is applied to an organic molecule, the protons either align with or against the magnetic field. Those aligned with the applied magnetic field are more stable and at a lower energy. Those aligned against the field are less stable and at a higher energy. When radio frequency, rf, of just the right frequency and energy, ΔE, interacts with the protons, those aligned with the field are excited to the higher energy state. This produces an absorption of energy that is detected by a ^{1}H NMR spectrometer.

The energy difference, ΔE, between these protons in a strong magnetic field is calculated as follows.

$$\Delta E = h\nu = \gamma \frac{h}{2\pi} H_o$$

In this equation, h is Planck's constant, H_0 is the strength of the applied magnetic field, γ is the gyromagnetic constant (26,753 s^{-1} gauss^{-1} for a proton), and ν is the frequency. The gyromagnetic constant depends on the magnetic moment of the nucleus.

The electrons that surround a H nucleus produce a magnetic field that opposes the applied magnetic field. Thus, the magnetic field at the nucleus is weaker than the applied magnetic field. The effect of the electrons around the nucleus is to shield them from the applied magnetic field, H_o. Hence, the radio frequency, rf, energy needed to excite a shielded nucleus to the higher energy state is greater than that of a "bare proton."

Protons in exactly the same electronic environment are termed equivalent protons. All equivalent protons absorb at the same frequency. If a proton is in a different electronic environment than another proton, they are called nonequivalent protons and they produce two different NMR signals.

The specific absorptions in the NMR spectrum give scientists valuable information about the structure of the molecule. The number of different absorptions tells how many different protons are in the molecule. The intensities of the absorptions are directly related to the number of protons. The position in the spectrum gives information about the electronic environment.

The position of a peak in an NMR spectrum is designated by a chemical shift. Chemical shifts, δ, are calculated by dividing the shift in Hz downfield from the TMS peak (tetramethyl silane, $(CH_3)_4Si$, peak) by the frequency of the spectrometer in MHz. Chemical shifts are measured in units of parts per million, ppm, downfield of the TMS peak which is at 0.0 ppm. The chemical shifts of selected protons are shown in the following table.

Type of Proton	Chemical Shift, ppm
H–C–R	0.9–1.8
H–C–C=C	1.6–2.6
H–C–C=O	2.2–2.5
H–C≡C–	2.5
H–C–C$_6$H$_5$	2.3–2.8
H–C=C–	4.5–6.5
H–C$_6$H$_5$	6.5–8.5
H–C=O	9–10
H–C–Cl	3.1–4.1
H–C–Br	2.7–4.1
H–C–O	3.3–3.7
H–C–NR	1–3
H–OR	0.5–5
H–OAr	6–8

NMR signals are split into more than one peak by protons on adjacent C atoms. This is called spin-spin splitting. These adjacent H atoms can either increase or decrease the shielding of a proton; hence, they can both increase and decrease the chemical shift of a proton. If only one H atom is on an adjacent C atom, then the peak is split into a doublet, a double peak. If two H atoms are on adjacent C atoms, then the peak is split into a triplet, a triple peak. In general, the signal is split into *n + 1* peaks, where *n* is the number of adjacent H atoms. For example, the spectrum of ethyl chloride, CH_3CH_2–Cl has two peaks. The methyl protons are split into a triplet because it has two adjacent protons, and the methylene protons are split into a quartet because they are adjacent to three protons.

Besides ^{1}H NMR spectroscopy, organic chemists also use ^{13}C NMR spectroscopy to help identify the structure of organic molecules. The ^{13}C NMR spectrum is generated in a manner similar to that of ^{1}H NMR spectrum, but is more difficult to obtain because only a small percent of C atoms are the heavier ^{13}C isotope. A technique called Fourier transform NMR is used to obtain these spectra.

Each nonequivalent C atom produces a signal in the ^{13}C NMR spectrum. For example, 1-bromobutane produces four peaks that correspond to the four C atoms in the chain. ^{13}C NMR spectra range more than 200 ppm from the TMS peak. The following table summarizes the ranges of chemical shifts in ^{13}C NMR spectra.

Type of Carbon Atom	Chemical Shift, ppm
RCH_3	0–35
R_2CH_2	15–40
R_3CH	25–50
RCH_2NH_2	35–50
RCH_2OH	50–65
$C\equiv C$	65–90
$C=C$	110–150
Aromatic	110–150
RCOOR	170–175
RCOOH	175–185
RCHO	190–220
RCOR′	204–220

In mass spectrometry, a sample is ionized and accelerated in a magnetic field toward an ion detector. During the process of ionization, a molecule is converted to an ion and at the same time the molecule is fragmented into characteristic parts that can be analyzed. The degree to which the molecular cation and cation fragments are deflected in the magnetic field depends on the mass-to-charge ratio, *m/e*. Ions with smaller *m/e* values are deflected more than those with higher values. By changing the magnetic field strength, ions with different *m/e* values can be focused on the mass spectrometer detector.

The mass spectrum of a compound is a bar graph that shows the number and intensities of the peaks produced plotted versus their masses. The highest peak in the mass spectrum is the base peak and is assigned the value of 100. The values for all other peaks are determined relative to the base peak. The peak that corresponds to the unfragmented molecule is called the parent or molecular ion peak. It has a value that corresponds to the molecular mass of the compound. Many mass spectra do not have a parent peak because the molecule totally fragments.

Test Yourself

1. Explain the difference between IR active and IR inactive vibrations. Give an example of each.

2. Explain how the IR spectrum of 1-pentyne differs from that of pentane.

3. Explain how the IR spectra of aldehydes differ from those of ketones.

    (For problems 4 and 5, consider the IR spectrum of Compound *A*.)

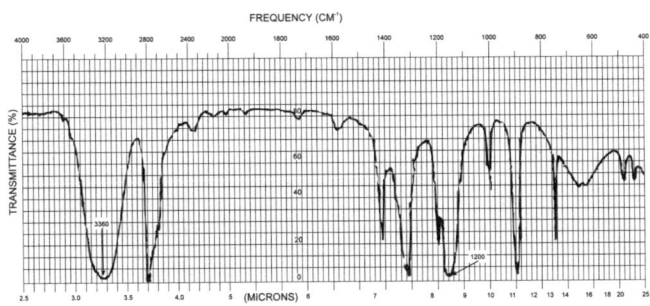

4. What functional group is present in compound *A* that gives the strong IR peaks at 1200 cm^{-1} and 3360 cm^{-1}?

5. In what general class of compounds (aromatic ketone, aliphatic acid, etc.) is *A*?

6. Compound *B* has the molecular formula of $C_9H_{10}O$. From the following IR spectrum, propose a structure of *B*.

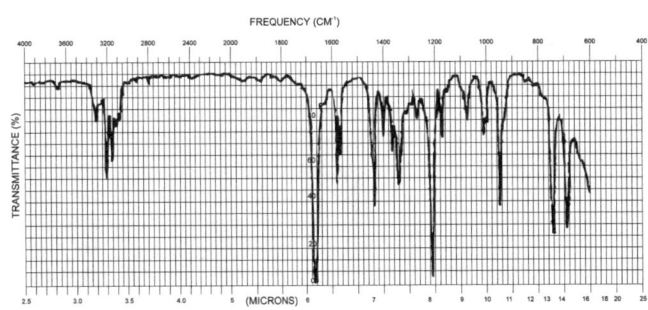

7. Consider the structure of methyl salicylate.

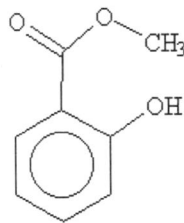

    Methyl salicylate

    The principal peaks in the IR spectrum of methyl salicylate are 3300 cm^{-1}, 2990 cm^{-1}, 1700 cm^{-1}, 1590 cm^{-1}, and 1540 cm^{-1}. Which bonds and/or structures produce these peaks?

8. Write a paragraph that explains how a signal is generated in ^1H NMR spectroscopy, also called proton magnetic resonance.

9. Explain the factors that influence the energy difference, ΔE, of protons in strong magnetic fields.

10. Completely describe the NMR spectrum of the following compound.
    $$Br-CH_2CH_2-O-CH_2CH_3$$

Do not forget to state how many signals are produced and how each is split.

11. Compound *C* has the molecular formula of $C_4H_6Cl_2$. Using the following ^1H NMR information, draw the structure of Compound *C* and explain its NMR spectrum.

Chemical shift, ppm	Splitting	Number of H atoms
2.2	singlet	3
4.1	doublet	2
5.7	triplet	1

12. Compound *D* has the molecular formula of C_4H_7NO. Its IR spectrum reveals a strong peak at 2240 cm^{-1} and a broad, strong peak at 3400 cm^{-1}. *D* has two absorptions in its ^1H NMR spectrum. A singlet at 1.65 ppm produced by six H atoms, and a singlet at 3.7 ppm produced by one H atom. Propose a structure for *D*.

13. Compound *E* is an ester that has the following NMR spectrum.

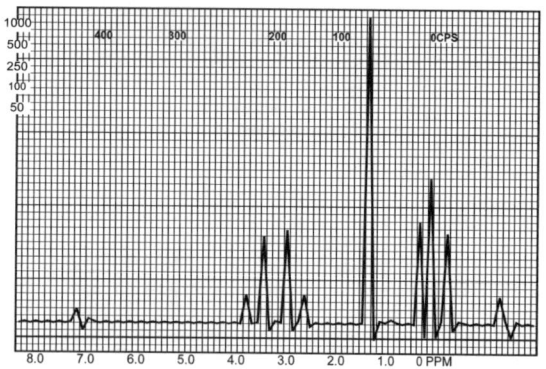

If the molecular formula of *E* is $C_4H_8O_2$, propose a structure consistent with the spectrum. Explain all peaks in the spectrum.

14. Discuss the problems and complications of using ^{13}C NMR spectroscopy.

15. Compound *F* has the molecular formula of $C_6H_4Cl_2$. Its ^1H NMR spectrum shows that all of the H atoms are equivalent. Its ^{13}C NMR spectrum shows two different types of C

atoms. Propose a structure of *F* consistent with the NMR data.

16. The ^1H NMR spectrum of cyclohexane shows one sharp peak at 1.4 ppm at room temperature. If the temperature of cyclohexane is lowered sufficiently, two broad peaks that split each other can be observed. Write a paragraph to explain the differences in these NMR spectra.

17. Explain what the off-resonance decoupled ^{13}C NMR spectrum of 1,2,2-trichloropropane would look like.

18. Consider the mass spectrum for compound *W*. If *W* is a hydrocarbon, propose a structure consistent with this mass spectrum.

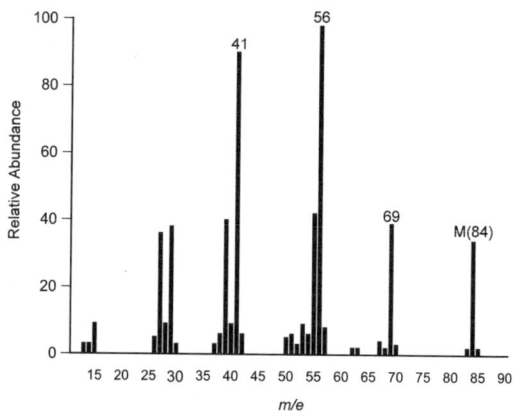

19. Compound *X* has the molecular formula of $C_{10}H_{14}O_2$. Its ^1H NMR spectrum has a six-hydrogen triplet at 1.0 ppm, a four-hydrogen quartet at 3.9 ppm, and a four-hydrogen singlet at 6.9 ppm. Propose a structure for *X*.

20. Which of the following corresponds to the frequency of infrared radiation?

    a. rotational energy states of molecules

    b. vibrational energy states of molecules

    c. π to π^* transitions in molecules

    d. nuclear energy states of molecules

21. Which of the following ranges of electromagnetic radiation is the fingerprint region in IR spectroscopy?

    a. 250 to 625 cm^{-1}

    b. 625 to 1300 cm^{-1}

    c. 2500 to 4000 cm^{-1}

    d. none of these

22. Which of the following bonds has the highest stretching frequency?

    a. C(sp^3)–H bond

    b. C(sp^2)–H bond

    c. C(sp)–H bond

    d. all the frequencies are the same

23. Which of the following types of double bonds vibrates at the highest frequency?

    a. isolated C–C double bond

    b. conjugated C–C double bond

    c. aromatic C–C double bond

    d. all vibrate at the same frequency

24. In what general IR frequency range is the N–H stretch?

    a. 1600–1700 cm^{-1}

    b. 2100–2200 cm^{-1}

    c. 2900–3000 cm^{-1}

    d. 3200–3500 cm^{-1}

25. Which of the following is an incorrect statement about NMR spectroscopy?

    a. The amount of shielding depends on the electronic environment near the proton.

    b. The intensities of the signals depend on the number of protons producing that signal.

    c. The number of different absorptions depends on the number of different types of protons in the molecule.

    d. The splitting of signals depends on the number of adjacent protons.

    e. All are correct

26. A 60.0 MHz NMR spectrometer records an absorbance at frequency of 500 Hz downfield of the TMS peak. What is the chemical shift for this absorbance?

    a. δ = 6.00 ppm

    b. δ = 8.33 ppm

    c. δ = 1.20 ppm

    d. cannot be calculated from this data

27. Which of the following methyl protons are least shielded in ^{1}H NMR spectroscopy?

    a. CH$_3$–I

    b. CH$_3$–Br

    c. CH$_3$–Cl

    d. CH$_3$–OH

    e. CH$_3$–F

28. How many signals are produced in the ^{1}H NMR spectrum of 2,2,3,3-tetrachlorobutane?

    a. none

    b. one

    c. two

    d. three

29. Which compound has a ^{1}H NMR spectrum consistent with the following?

Chemical shift, ppm	Splitting	Number of H atoms
2.0	quintet	2
2.8	singlet	1
3.7	triplet	2
3.8	triplet	2

    a. ClCH$_2$CH$_2$CH$_2$OH

    b. CH$_3$CH$_2$CH$_2$OH

    c. HOCH$_2$CH$_2$CH$_2$OH

    d. H$_2$C=CHCH$_2$CH$_2$Cl

30. How many ^{1}H NMR signals (ignoring signal splitting) are in the spectrum of 1,1-dimethylcyclopropane?

    a. two

    b. three

    c. four

    d. five

e. none of these

31. Which of the following compounds produces a ^{13}C NMR spectrum with five signals, two between 10 and 30 ppm and the remaining between 120 and 140 ppm?

    a. ethylbenzene

    b. *o*-diethylbenzene

    c. *m*-diethylbenzene

    d. *p*-diethylbenzene

32. Consider the following two compounds.

    Which spectroscopic methods could **not** be used to readily distinguish these compounds?

    a. IR spectroscopy

    b. mass spectroscopy

    c. ^1H NMR

    d. ^{13}C NMR

    e. All of the above could be used to distinguish these compounds

33. Which of the following compounds exhibit spin-spin coupling in their ^1H NMR spectrum?

    a. *cis*-1,2-dichloroethene

    b. 1,2-dichloroethane

    c. 1-chloro-2,2-dimethylpropane

    d. 1-bromo-1-iodoethene

    e. none of these

34. One of the mass spectrum peaks, *m/e*, in hexane is 42. Which of the following structural fragments produces this peak?

    a. $CH_2CH_2CH_2^+$

    b. $CH_3CH_2CH_2^+$

    c. $CH_3CH_2CH_3^+$

    d. none of these

35. A compound has the formula of $C_{10}H_{12}O$. It has *m/e* peaks at 15, 43, 57, 91, 105, and 148. Which of the following molecules produces this mass spectrum?

    a. $CH_3-O-CH_2CH_2CH_2CH_2CH_2C(CH_3)_3$

    b. $CH_3COCH_2CH_2-C_6H_5$

    c. $C_6H_5-COCH_2CH_2CH_3$

    d. none of these

✔ Check Yourself

1. To produce an IR absorption, a vibration of a bond must change its dipole moment. This is called IR active. Vibrations that do not change the dipole moment are IR inactive. The C–C double bond stretch in 1-propene is an IR active vibration. The C–C double bond stretch in ethene is an IR inactive vibration. **(IR theory)**

2. The molecules 1-pentyne, $CH_3CH_2CH_2C≡CH$, and pentane, $CH_3CH_2CH_2CH_2CH_3$, differ at the C-C triple bond. Both molecules exhibit C–H stretches just below 3000 cm^{-1} and C–H bending around 1400 cm^{-1}. The IR spectrum of 1-pentyne additionally has a H–C≡ stretch around 3300 cm^{-1} and a C≡C stretch around 2100 cm^{-1}. **(IR spectra)**

3. The spectra of aldehydes and ketones both have the strong carbonyl stretches, C=O, near 1720 cm^{-1}. They differ because an aldehyde produces a C–H stretch because it has a H atom bonded to a carbonyl. This C–H stretch to the carbonyl is lower in frequency than the C–H stretch when the C atom is sp^3 hybridized. **(IR spectra)**

4. The strong peak at 1200 cm^{-1} is characteristic of the C–O stretch in alcohols. The broad peak from 3200 to 3500 cm^{-1} centered at 3360 cm^{-1} is the O–H stretch. These two strong peaks are found in alcohols. **(IR interpretation)**

5. This is the spectrum of an aliphatic alcohol. The spectrum does not show any aromatic absorptions. Because the carbonyl stretch is not present, it eliminates carbonyl compounds. No double or triple bond stretches are evident. The strong absorption at 1375 cm^{-1} is characteristic of methyl groups ("umbrella" peak). Actually, this is the IR spectrum of *t*-butyl alcohol. **(IR interpretation)**

6. This is the spectrum of 1-phenyl-1-propanone, C$_6$H$_5$–COCH$_2$CH$_3$. The IR spectrum shows the strong carbonyl absorbance at 1690 cm^{-1}. The two moderately strong peaks at 700 and 750 cm^{-1} show the presence of a monosubstituted aromatic. The strong peak at 1210 cm^{-1} is due to the C–CO–C group of a ketone. The absence of a broad peak above 3100 cm^{-1} discounts the possibility of an alcohol. **(IR interpretation)**

7. The peak at 3300 cm^{-1} is produced by the O–H stretch. It is usually a broad peak that results from H bonding. The 2990 cm^{-1} peak is the C–H stretch of the CH$_3$ group. The 1700 cm^{-1} peak is the carbonyl stretch, and the remaining peaks, 1590 cm^{-1} and 1540 cm^{-1}, are the result of the aromatic ring. **(IR interpretation)**

8. In the presence of a large magnetic field some of the proton nuclei align with the field and others against the field. Those aligned with the applied magnetic field are more stable and at a lower energy. Those aligned against the field are less stable and at a higher energy. When radio frequency, rf, of just the right frequency and energy interacts with the protons, those aligned with the field are excited to the higher energy state. This produces an absorbance of energy. **(NMR absorbance)**

9. The energy difference, ΔE, between these protons is calculated as follows.

$$\Delta E = h\nu = \gamma \frac{h}{2\pi} H_o$$

In this equation, h is Planck's constant, H_o is the strength of the applied magnetic field, γ is the gyromagnetic constant (26,753 s^{-1} gauss^{-1} for a proton), ν is the frequency. The gyromagnetic constant depends on the magnetic moment of the nucleus. **(NMR theory)**

10. This compound has four nonequivalent sets of protons. The four signals are split as follows. **(NMR spectra)**

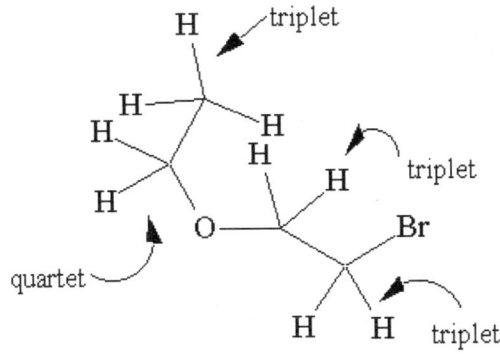

11. The compound that produced this NMR spectrum is 1,3-dichloro-2-butene.

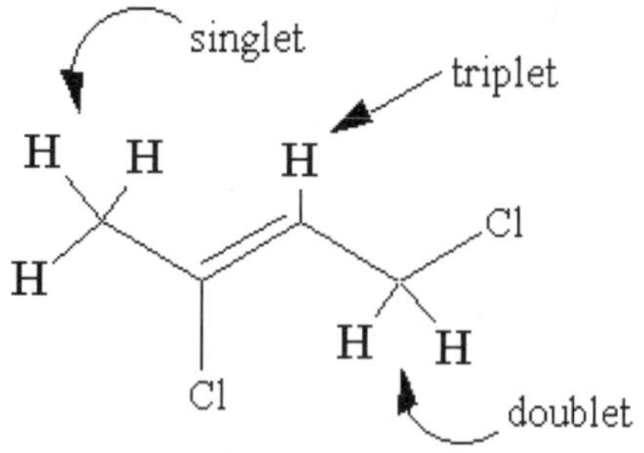

(NMR spectra)

12. The structure of Compound *D* is as follows.

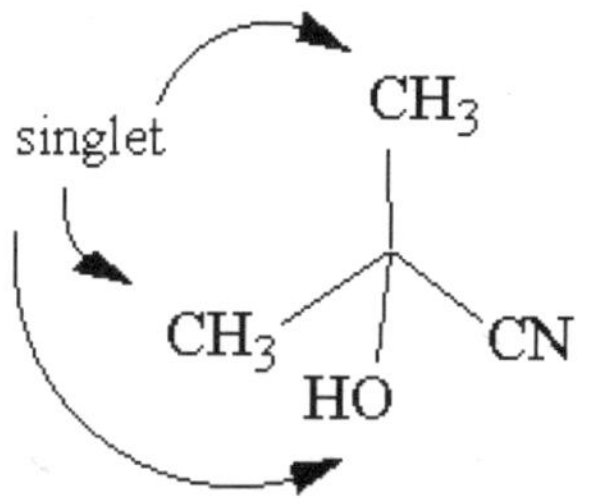

Each of the signals in the spectrum is a singlet because there are no H atoms on adjacent C atoms. The singlet at 1.65 ppm is produced by the six methyl H atoms and the singlet at 3.7 ppm is produced by the proton bonded to the O atom. The 2240 cm^{-1} peak in the IR spectrum is produced by the nitrile, and the broad peak at 3400 cm^{-1} is produced by the OH group. **(NMR and IR spectra)**

13. Compound *E* is ethyl ethanoate, $CH_3COOCH_2CH_3$. The unsplit singlet at 2.1 ppm is produced by the methyl group bonded to the C atom of the carbonyl group. The remaining triplet at 1.3 ppm (CH_3) and quartet at 4.2 ppm (CH_2) are from the H atoms in the ethyl group. The methylene H atoms are farther downfield because they are bonded to an O atom. **(NMR interpretation)**

14. ^{13}C NMR spectroscopy is more difficult than ^1H NMR because of the low percent abundance of the ^{13}C isotope. Additionally, ^{13}C is less sensitive than protons to the process of NMR. Complications are produced with $^{13}C-^1H$ and $^{13}C-^{13}C$ coupling. $^{13}C-^1H$ coupling can be eliminated by broad-band decoupling of the signals. As a result, only singlets are observed in ^{13}C NMR spectra. $^{13}C-^{13}C$ coupling is of little consequence because of the low abundance. (**^{13}C NMR spectra**)

15. *p*-Dichlorobenzene has the formula $C_6H_4Cl_2$.

p-dichlorobenzene

Its ^1H NMR spectrum has all equivalent H atoms because they are in the same environment. Its ^{13}C NMR spectrum has two different types of C atoms. **(NMR spectra)**

16. At 298 K, the interconversion of the chair forms of cyclohexane is rapid compared to the time scale of ^{1}H NMR.

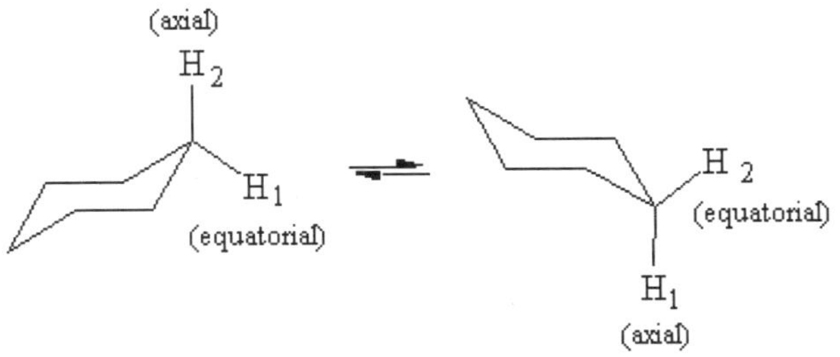

Hence, only one peak is observed. At low temperatures, the interconversion is slowed and the ^{1}H NMR can detect the difference between the nonequivalent axial and equatorial H atoms. Thus, two signals are produced that split each other. (**NMR interpretation**)

17. Off-resonance decoupling allows the protons to split the NMR signals of C atoms to which they are bonded. The structure of 1,2,2-trichloropropane is as follows.

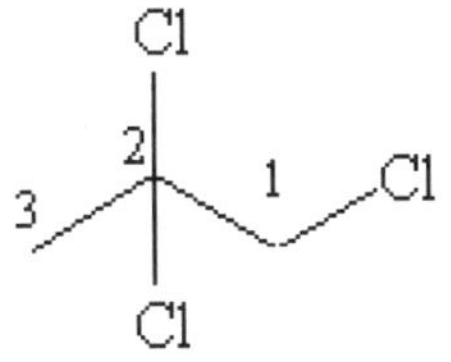

1,2,2-trichloropropane

Because the three C atoms are nonequivalent, it produces three signals in its ^{13}C NMR spectrum. The second C atom will be downfield farther than the other two because of the electron-withdrawing effects of the two Cl atoms. This peak should be near 87 ppm. Because there are no H atoms bonded to it, the peak will be a singlet. The next peak upfield will be the first C atom because it is bonded to one Cl atom. The peak should be near 56 ppm and will be a triplet because of the splitting by the two H atoms. The final peak, produced by the third C atom, will be a quartet near 33 ppm. (**^{13}C NMR**)

18. Compound *W* is 2-methyl-1-pentene.

The molecular ion, $m/e = 84$, means the molecular formula is C_6H_{12}. The removal of a CH_3 ($m/e = 15$) gives the peak at 69. The production of the allylic cation produces a m/e peak at 41. This peak is characteristic of most alkenes. The base peak with m/e of 56 corresponds to $CH_3CH_2CH=CH_2$. (**Mass spectroscopy**)

19. Structure *X* is 1,4-diethoxybenzene.

1,4-diethoxybenzene

The four H atoms on the aromatic ring produce the singlet at 6.9 ppm. The four methylene H atoms bonded to the O atom produce the quartet at 3.9 ppm. The six methyl H atoms produce a triplet at 1.0 ppm. (**NMR interpretation**)

20. b. Infrared radiation excites molecules from their ground vibrational energy states to excited vibrational energy states. (**IR theory**)

21. b. The fingerprint region of the IR spectrum is from 625 to 1300 cm^{-1}. This region is characteristic of larger segments of the molecule rather than vibrations of individual bonds. (**IR spectra**)

22. c. The stretching frequency for the C(sp)-H bond is the highest because the higher s character makes a "tighter" bond that vibrates at a higher frequency. (**IR theory**)

23. a. Isolated C–C double bonds vibrate at the highest frequency (1640–1680 cm^{-1}). Conjugation decreases the electron density between the C atoms and lowers the vibrational frequency. (**IR theory**)

24. d. The N–H stretch in amines and amides is generally from 3200 to 3500 cm^{-1}. These peaks can be either sharp or broad. (**IR spectra**)

25. e. All are correct statements about NMR spectroscopy. (**NMR spectroscopy**)

26. b. The chemical shift, δ, is 8.33 ppm (δ = 500 Hz/60.0 MHz). Chemical shifts are calculated by dividing the shift in Hz downfield from the TMS peak by the frequency of the spectrometer in MHz. (**Chemical shifts**)

27. e. The methyl protons in CH$_3$–F are most deshielded (δ = 4.3 ppm) because the F atom is most electronegative and thus pulls the most electron density away from the protons. (**Chemical shifts**)

28. b. 2,2,3,3-Tetrachlorobutane only produce one signal because all of the protons are equivalent. (**NMR spectra**)

29. a. The only structure with these chemical shifts and splitting patterns is ClCH$_2$CH$_2$CH$_2$OH, 3-chloro-1-propanol. The two H atoms on the second C atom produce a quintet at 2.0 ppm. The two H atoms on the first C atom yield the triplet at 3.7 ppm. The singlet at 2.8 ppm is produced by the proton bonded to the methyl group. Finally, the two H atoms on the third C atom produce a triplet at 3.8 ppm. (**NMR spectra**)

30.   a. 1,1-Dimethylcyclopropane has two signals. Its structure is as follows.

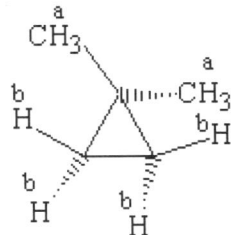

The six methyl H atoms produce one signal (a). The remaining four protons are equivalent and produce the second signal (b). (**NMR interpretation**)

31.   b. *o*-Diethylbenzene produces five signals in its ^{13}C NMR spectrum. The two C atoms in the equivalent ethyl groups produce the two signals in the 10 to 30 ppm range. Three different signals are produced from the benzene ring: the two C atoms bonded to the ethyl groups, the two C atoms *ortho* to the ethyl groups, and those *meta* to the ethyl groups. (**^{13}C NMR**)

32.   e. All of the spectroscopic methods give different spectra that could be used to distinguish them. (**Spectroscopy**)

33.   d. The H atoms in 1-bromo-1-iodoethene are adjacent to different halogen atoms and thus are nonequivalent. Each of these H atoms produces a doublet. (**NMR coupling**)

34.   a. The fragment $CH_2CH_2CH_2{}^+$ is the only one with a mass of 42. It would be produced from the three interior C atoms of *n*-hexane. (**Mass spectroscopy**)

35.   b. $CH_3COCH_2CH_2–C_6H_5$ is the molecule that produces the given mass spectrum. The peak at 15 is produced by the $CH_3{}^+$ group. A peak at 43 would suggest an acetyl group, CH_3CO group. The peak at 57 would be the result of adding a CH_2 to the acetyl group, CH_3COCH_2. When an acetyl group separates, *m/e* = 43, from the parent molecule, which has a *m/e* of 148, it gives a value of 105. Removing a CH_3COCH_2 group from the parent gives the 91 peak. (**Mass spectroscopy**)

Grade Yourself

Circle the number of questions you missed, then fill in the total incorrect for each topic. If you answered more than three questions incorrectly, you need to focus on that topic. (If a topic has less than three questions and you had at least one wrong, we suggest you study that topic also. Read your textbook, a review book, or ask your teacher for help.)

Subject: IR Spectroscopy, NMR Spectroscopy, and Mass Spectrometry

Topic	Question Numbers	Number Incorrect
IR theory	1, 20, 22, 23	
IR spectra	2, 3, 21, 24	
IR interpretation	4, 5, 6, 7	
NMR absorbance	8	
NMR theory	9	
NMR spectra	15, 28, 29	
NMR interpretation	16, 19, 30	
13C NMR	17, 31	
Mass spectroscopy	18, 34, 35	
NMR spectroscopy	25	
Chemical shifts	26, 27	
Spectroscopy	32	
NMR coupling	33	

Alcohols and Phenols

13

Brief Yourself

Alcohols, ROH, can be prepared in many ways. Alkenes undergo acid-catalyzed hydration to produce alcohols.

$$R_2C=CH_2 + H_2O \rightarrow R_2CH(OH)CH_3$$

Oxymercuration-demercuration, another way to hydrate alcohols, can be used to get higher yields of alcohols. In this reaction, the alkene first reacts with mercury(II) acetate, $Hg(OAc)_2$, in THF and water. Then, the product is reduced with sodium borohydride, $NaBH_4$. To get the anti-Markovnikov product in the hydration of an alkene, hydroboration-oxidation is used.

$$R_2C=CH_2 + H_2O \rightarrow R_2CHCH_2OH$$

In this procedure, the alcohol is first treated with diborane, B_2H_6, and the resulting product is oxidized with hydrogen peroxide and base. Another way to produce alcohols is to hydrolyze alkyl halides.

$$RX + OH^- \rightarrow ROH + X^-$$

This reaction works well as long as the alkyl halide does not readily undergo E2 elimination.

Organometallic compounds can also be used to synthesize alcohols. Grignard reagents, RMgX, and organolithium compounds, RLi, react with aldehydes and ketones and produce alcohols. Both Grignard reagents and organolithium compounds react with formaldehyde followed by hydrolysis to give primary alcohols.

$$RMgX (RLi) + H_2C=O \rightarrow RCH_2OH$$

These organometallic compounds react with aldehydes to give secondary alcohols and ketones to produce tertiary alcohols. Finally, they also react with esters and produce tertiary alcohols.

Alcohols can also be produced by the reduction of aldehydes and ketones. Aldehydes are reduced to primary alcohols using H_2 on Pt, Pd, or Ni catalysts.

$$RCHO + H_2 \text{ (cat.)} \rightarrow RCH_2OH$$

Ketones reduce under the same conditions to secondary alcohols. In addition to using H_2 and a metal catalyst, sodium borohydride, $NaBH_4$, and lithium aluminum hydride, $LiAlH_4$, may also be used. Neither $NaBH_4$ nor $LiAlH_4$ reduce C–C double bonds. $LiAlH_4$ can also be used to reduce carboxylic acids to primary alcohols. When esters are reduced, they produce two alcohols.

$$RCO_2R' + H_2 \text{ (cat.)} \rightarrow RCH_2OH + R'OH$$

Primary alcohols can also be produced from epoxides. This is accomplished by reacting a Grignard reagent with ethylene oxide in ether followed by acid hydrolysis. The resulting alcohol has two more C atoms than the Grignard reagent.

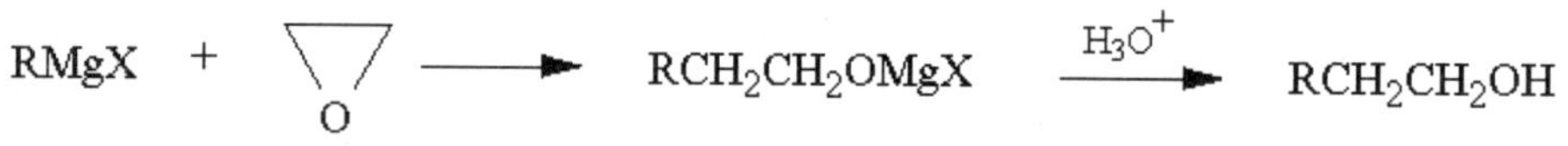

Alcohols are reactive with many compounds. For example, sodium and other alkali metals react with alcohols producing alkoxides.

$$2ROH + 2Na \rightarrow 2RO^- Na^+ + H_2$$

Alcohols react with hydrogen halides, HX, to produce alkyl halides

$$ROH + HX \rightarrow RX + H_2O$$

Alcohols react with thionyl chloride, $SOCl_2$, and produce alkyl chlorides. They also react with phosphorus trihalides, PX_3, and yield alkyl halides.

Alcohols undergo acid-catalyzed dehydration to alkenes.

$$RCH_2CH_2OH + H_2SO_4 \text{ (heat)} \rightarrow RCH=CH_2 + H_2O$$

Two alcohols may condense to an ether if heated in strong acid.

$$2ROH + H_2SO_4 \text{ (heat)} \rightarrow ROR + H_2O$$

Alcohols react with carboxylic acids, R'COOH, in the presence of an acid catalyst and produce esters.

$$ROH + R'COOH \rightarrow R'COOR + H_2O$$

Alcohols can also form inorganic esters with acids such as H_2SO_4 or H_3PO_4.

Primary and secondary alcohols can be oxidized to carbonyl compounds. Primary alcohols first oxidize to aldehydes and then to carboxylic acids using chromic acid or permanganate.

$$RCH_2OH \rightarrow RCHO \rightarrow RCOOH$$

Secondary alcohols oxidize to ketones under the same conditions. Tertiary alcohols cannot oxidize to carbonyl compounds because they lack a H atom on the C atom bearing the OH group. Both the Collins $((C_5H_5N)_2CrO_3$ in dichloromethane) and PCC (pyridinium chlorochromate in dichloromethane) reagents can be used to oxidize primary alcohols only to aldehydes.

Phenols are compounds that have an OH group bonded to an aromatic ring. The simplest phenol is C_6H_5–OH, called phenol. Phenols are significantly more acidic than nonaromatic alcohols because of the resonance stabilization of the resulting phenoxide ion, C_6H_5–O^- (the conjugate base). Any substitutents that help disperse the negative charge in the phenoxide ion increase the acidity of phenols. For example, *p*-nitrophenol is about 1000 times more acidic than phenol. The OH group is a strongly activating group on an aromatic ring and is an *ortho-para* director. Phenols undergo electrophilic aromatic substitution reactions. Care must be taken to prevent multiple substitutions because the OH group strongly activates the ring. Phenols undergo many of the same reactions as other benzene derivatives.

Test Yourself

1. How could a primary alcohol be synthesized using a Grignard reagent? Write a general equation that shows this synthesis.

2. a. Draw the structure of the alcohol that results when ethyl magnesium bromide reacts with cyclohexanone in ether, followed by acid hydrolysis.

   b. What general type of alcohol forms?

3. Draw the structure and write the name of the product that results when vinyl lithium, $H_2C=CHLi$, reacts with benzophenone, C_6H_5–CO–C_6H_5, in ether, followed by acid hydrolysis.

4. List three sets of reactants that could be used in a Grignard synthesis of 2-cyclopentyl-2-pentanol.

5. What ester reacts with ethyl magnesium chloride to produce the following alcohol?

6. List seven different reactions that could be used to synthesize alcohols.

7. What alcohol results when cycloheptanone reacts first with lithium aluminum hydride in ether followed by hydrolysis?

8. What diol results when 1,2-diethylcyclohexene reacts with a cold solution of potassium permanganate? Do not forget stereochemistry.

9. a. What type of reaction takes place when primary alcohols are heated with strong acids?

    b. Draw the structure of the product that results when 2,4-dimethyl-1,5-pentanediol is heated in sulfuric acid.

10. List three reagents that could be used to convert alcohols to alkyl halides.

11. What is the product of the reaction of ethanol with the following compound using an acid catalyst?

12. Outline a synthetic pathway to convert propanal to acetone, propanone.

13. Write the products that result when excess ethanol reacts with phosphoric acid.

14. Draw the structure of the product that results when cyclooctanol reacts with *p*-toluenesulfonyl chloride in pyridine.

15. Outline a synthetic pathway for the following conversion.

16. Explain why phenol is about one million times more acidic than ethanol.

17. Write an equation that shows the oxidation of 1,4-benzenediol, hydroquinone, using chromic acid.

18. Starting with phenol and all necessary reagents show how phenol could be converted to *n*-propyl phenyl ether.

19. Starting with $CH_3CH_2CH_2COCl$ and all necessary reagents show how 2-methylpentane is synthesized.

20. Which of the following results when methyl benzoate, $C_6H_5-COOCH_3$, reacts with $LiAlH_4$ followed by hydrolysis?

    a. methanol

    b. benzyl alcohol

    c. ethanol

    d. phenol

    e. a and b are correct

    f. c and d are correct

21. What product(s) results when *n*-pentyl magnesium bromide reacts with ethylene oxide (oxirane) followed by acid hydrolysis?

    a. 1-pentanol

    b. 1-hexanol

    c. 1-heptanol

    d. none of the above

22. Which of the following best describes the reaction of a Grignard reagent and a carbonyl compound?

    a. The partially positive C atom in the Grignard reagent attacks the partially negative carbonyl C atom.

    b. The partially negative C atom in the Grignard reagent attacks the partially positive carbonyl C atom.

    c. The oxygen atom of the carbonyl group attacks the negative C atom in the Grignard reagent.

    d. none of these

23. 1-Butene reacts with hydrogen bromide and produces Compound *A*, which reacts with magnesium in ether and produces Compound *B*. *B* is first treated with formaldehyde and the product is hydrolyzed to produce Compound *C*. What is the identity of *C*?

    a. 1-pentanol

    b. 2-methyl-1-butanol

    c. 3-methyl-1-butanol

    d. 2-butanol

    e. none of these

24. Which of the following will react with excess methyl magnesium bromide followed by acid hydrolysis to produce 2,3,3-trimethyl-2-pentanol?

    a. $CH_3C(CH_3)_2COOCH_3$

    b. $CH_3CH_2C(CH_3)_2CHO$

    c. $CH_3CH_2C(CH_3)_2COOH$

    d. none of these

25. Which of the following sets of reagents make the following conversion?

    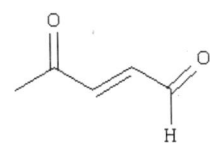

    a. 1. D_3O^+, 2. $NaBH_4/OH^-$, 3. H_3O^+

    b. 1. $NaBD_4$, 2. H_3O^+

    c. D_2/Pt

    d. 1. $KMnO_4/OH^-$, 2. $NaBD_4$, 3. H_2O

26. Which of the following reacts with cyclopentanol to produce the following compound?

    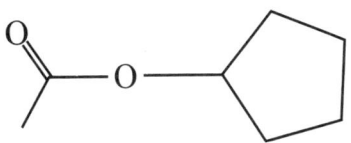

    a. $CH_3COOCOCH_3$

    b. CH_3COCl

    c. CH_3COOH/H^+

    d. all of these

    e. none of these

27. What is the product of the reaction of 1-methylcyclopentanol with sodium dichromate and acetic acid?

    a. cyclopentanone

    b. cyclopentylmethanal

    c. $C_5H_9–CO–COOH$

    d. no reaction

28. What product results when the following compound is treated with excess $LiAlH_4$ followed by hydrolysis?

    a. $CH_3COCH=CHCH_2OH$

    b. $CH_3CH(OH)CH=CHCH_2OH$

    c. $CH_3CH(OH)CH_2CH_2CH_2OH$

    d. $CH_3COCH_2CH_2CHO$

29. Which of the following is the strongest acid?

    a. phenol

    b. 2,4-dinitrophenol

    c. 3,5-dinitrophenol

    d. 2,4,6-trinitrophenol

30. What is the major product that results when phenol reacts with excess Br_2 and a strong Lewis acid catalyst?

    a. *o*-bromophenol

    b. *m*-bromophenol

    c. *p*-bromophenol

    d. 2,4,6-tribromophenol

31. What product results when *m*-methylphenol, *m*-cresol, reacts with two moles of *t*-butyl chloride and $AlCl_3$?

    a. 4,6-di-*t*-butyl-4-methylphenol

    b. 2,4-di-*t*-butyl-3-methylphenol

    c. 2,4-di-*t*-butyl-5-methylphenol

    d. none of these

32. 2-Butanol is oxidized by sodium dichromate and sulfuric acid in acetone to Compound *D*. Another sample of 2-butanol is reacted with PBr_3 producing Compound *E*. *E* is heated with Mg and ether to produce Compound *F*. *D* and *F* are reacted and the product is hydrolyzed to *G*. What is the identity of *G*?

    a. 3,4-dimethyl-3-hexanol

    b. 3,4-dimethyl-2-hexanol

    c. 2-octanol

    d. none of these

33. Which of the following reagents oxidize 1-propanol to propanal?

    a. pyridinium chlorochromate (PCC reagent)

    b. $CrO_3 \cdot 2$ pyridine (Collins reagent)

    c. $K_2Cr_2O_7$, H_2SO_4, acetone (Jones reagent)

    d. a and b

    e. b and c

    f. none of these

34. Which of the following react most rapidly when mixed with the Lucas reagent, $ZnCl_2$ and HCl?

    a. primary alcohols

    b. secondary alcohols

    c. tertiary alcohols

    d. all react at the same rate

35. Consider the following reaction.

    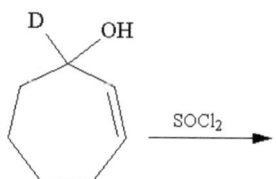

    Compound I, II, III, and IV are possible products of the reactions.

    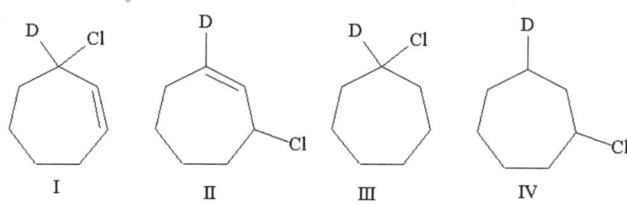

    Which of the following is the product(s) of this reaction?

    a. I

    b. II

    c. III

    d. IV

    e. a and b

    f. c and d

Check Yourself

1. A primary alcohol can be synthesized by reacting a Grignard reagent, RMgX, with formaldehyde, $H_2C=O$, typically in ether. The resulting primary alkoxymagnesium halide is then hydrolyzed. **(Grignard synthesis of alcohols)**

$$RMgX + H_2C=O \rightarrow RCH_2-O-MgX$$

$$RCH_2-O-MgX + H^+ \rightarrow RCH_2\text{-}OH$$

2. a. The product of the reaction is 1-ethylcyclohexanol.

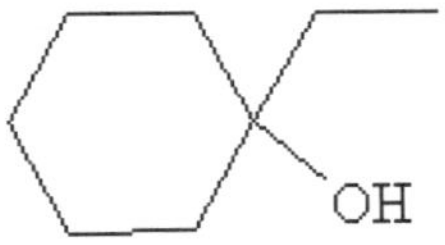

1-ethylcyclohexanol

b. This is a tertiary alcohol. **(Grignard synthesis of alcohols)**

3. The product of this reaction is as 1,1-diphenyl-2-propen-1-ol.

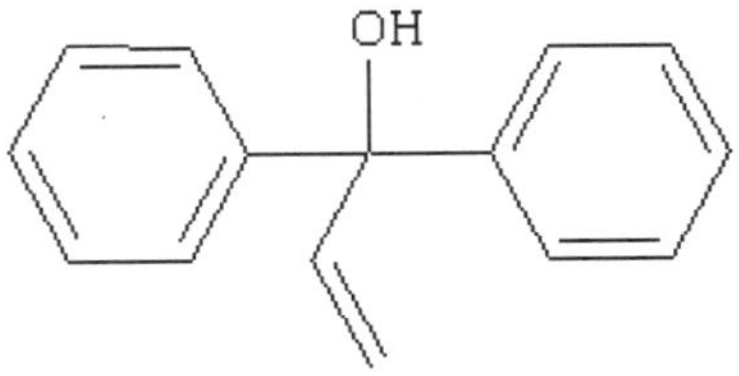

1,1-diphenyl-2-propen-1-ol

Organolithium syntheses of alcohols are similar to Grignard syntheses. **(Organolithium synthesis of alcohols)**

4. 1. $CH_3MgBr + C_5H_9\text{-}CO\text{-}CH_2CH_2CH_3 \rightarrow$
   2. $C_5H_9MgBr + CH_3\text{-}CO\text{-}CH_2CH_2CH_3 \rightarrow$
   3. $CH_3CH_2CH_2MgBr + C_5H_9\text{-}CO\text{-}CH_3 \rightarrow$
   (Grignard synthesis of alcohols)

5. An ester that reacts with ethyl magnesium bromide and produces 3-ethyl-5-methyl-3-hexanol is as follows.

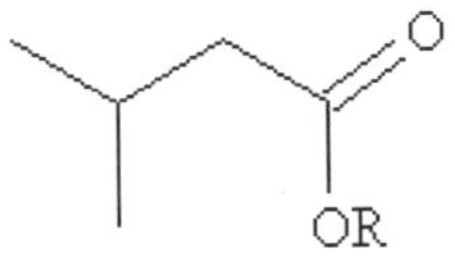

Esters are used to produce tertiary alcohols in Grignard syntheses. **(Grignard synthesis of alcohols)**

6. 1. acid-catalyzed hydration of alkenes
   2. oxymercuration-demercuration of alkenes
   3. hydroboration-oxidation of alkenes
   4. nucleophilic substitution of alkyl halides and tosylates
   5. Grignard synthesis using aldehydes and ketones
   6. organolithium synthesis using aldehydes and ketones
   7. reduction of carbonyl compounds
   (Alcohol synthesis)

7. The product of the reaction of cycloheptanone and lithium aluminum hydride is cycloheptanol. **(Alcohol synthesis by carbonyl reduction)**

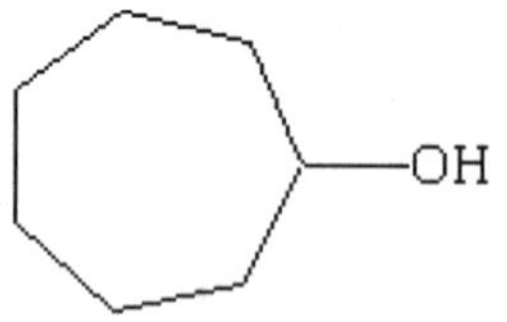

8. The reaction of a cold solution of permanganate produces vicinal diols in which the two OH groups are on the same side of the molecule as a result of a syn addition. Thus, the product of this reaction is the *cis* isomer of 1,2-diethyl-1,2-cyclohexanediol. **(Diol synthesis)**

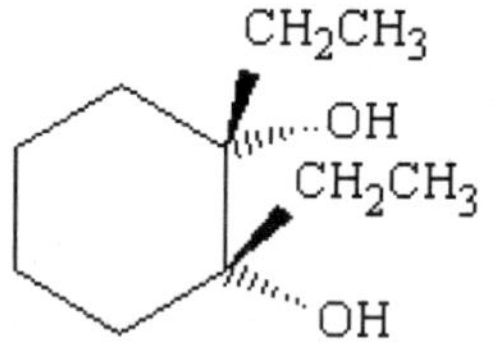

9. a. Primary alcohols form ethers when heated with strong acids such as H_2SO_4. This is called a condensation reaction because the product is produced with the elimination of water or another small molecule.

   b. The product of this reaction is as follows. **(Alcohol condensation reactions)**

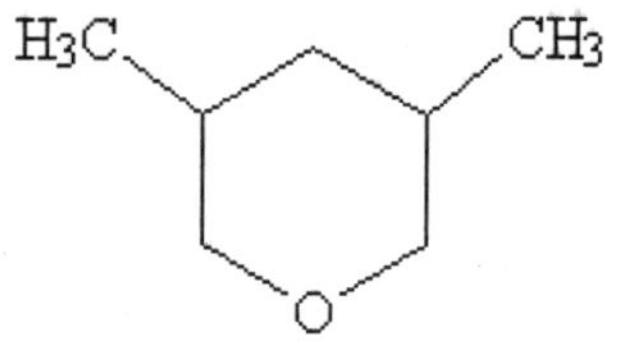

10. 1. HX, hydrogen halides
    2. $SOCl_2$, thionyl chloride
    3. PX_3, phosphorus trihalides
    (Converting alcohols to alkyl halides)

11. Alcohols react with carboxylic acids, for this problem a diacid, and produce esters. The product of this reaction is diethylmalonate. **(Esterification)**

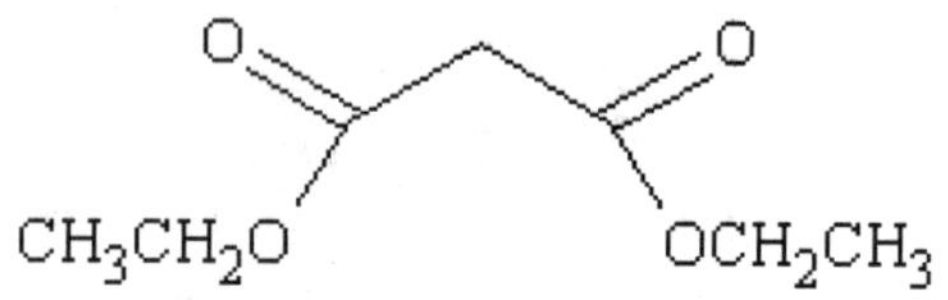

12. Begin by reducing propanal to 1-propanol using LiAlH4 in ether followed by hydrolysis. Next, dehydrate 1-propanol to propene using sulfuric acid. Hydrate the propene to 2-propanol using oxymercuration-demercuration, $H_2O/Hg(OAc)_2$ followed by $NaBH_4$. Finally, oxidize the 2-propanol to acetone using chromic acid. (**Alcohol synthesis**)

13. Alcohols react with phosphoric acid and produce alkyl dihydrogen phosphates, dialkyl hydrogen phosphates, and trialkyl phosphates. Thus, the products of the reaction of ethanol with H_3PO_4 are $CH_3CH_2H_2PO_4$, $(CH_3CH_2)_2HPO_4$, and $(CH_3CH_2)_3PO_4$. Additionally, if the conditions are right, phosphoric acid can dehydrate ethanol to diethylether. (**Alcohol reactions**)

14. The product of the reaction is cyclooctyl *p*-toluenesulfonate. (**Formation of sulfonate esters**)

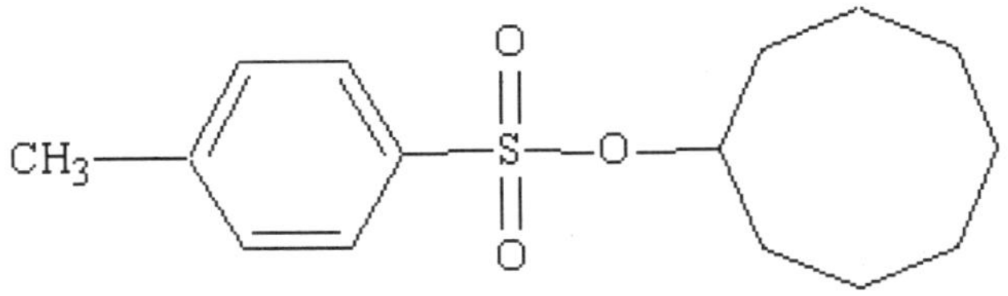

15. The conversion is accomplished as follows. (**Alcohol synthesis**)

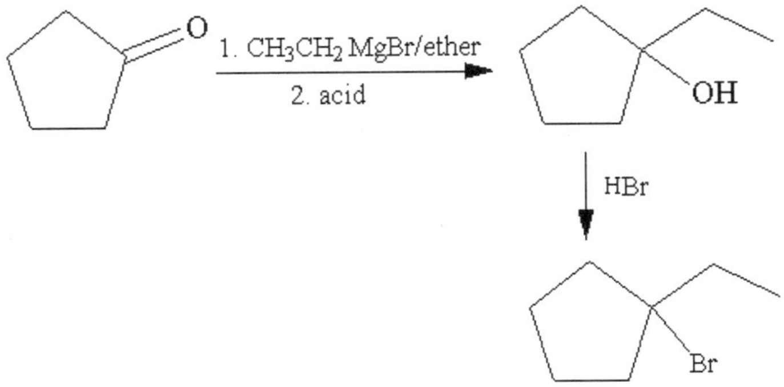

16. Phenol, C_6H_5–OH, is significantly more acidic than ethanol because the resulting phenoxide ion, C_6H_5–O⁻, stabilizes the negative charge through resonance, electron delocalization. Electron delocalization is not possible when ethanol produces the ethoxide ion, CH_3CH_2–O⁻. (**Phenol properties**)

17. The oxidation of hydroquinone by chromic acid produces *p*-benzoquinone. (**Phenol reactions**)

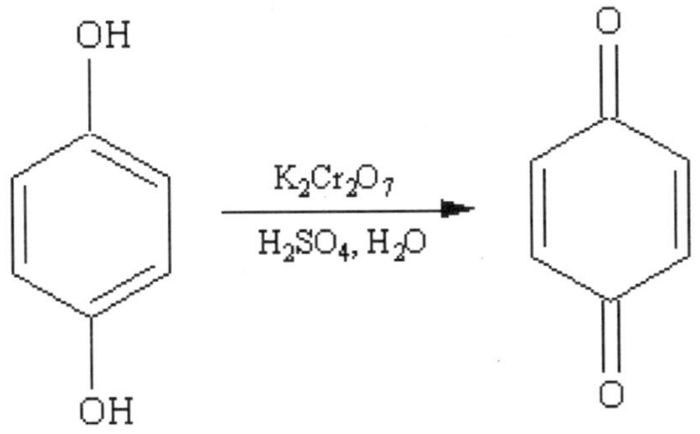

1,4-benzenediol   p-benzoquinone

18. Because phenol is acidic, it will react with NaOH and produce sodium phenoxide, C_6H_5–ONa. Sodium phenoxide, a source of the phenoxide ion, will displace Br in *n*-propyl bromide, a nucleophilic substitution reaction, producing *n*-propyl phenyl ether ($CH_3CH_2CH_2$–O–C_6H_5). **(Phenol reactions)**

19. 2-Methylpentane can be synthesized as follows.

    (Alcohol synthesis)

20. e. Both methanol and benzyl alcohol, C_6H_5–CH_2OH, result because $LiAlH_4$ reduces esters to two alcohols. **(Alcohol synthesis by carbonyl reduction)**

21. c. 1-Heptanol results because a Grignard reagent reacts with ethylene oxide and produces an alcohol with two additional C atoms. **(Grignard synthesis of alcohols)**

22. b. The partially negative C atom in the Grignard reagent attacks the partially positive carbonyl C atom. The C atom directly bonded to the Mg is partially negative because it is more electronegative than the metal atom. It bonds to the partially positive carbonyl C atom while the MgX bonds to the negative O atom. **(Grignard synthesis of alcohols)**

23. b. 2-Methyl-1-butanol is the identity of *C*. 1-Butene reacts with HBr and produces 2-bromobutane *(A)*, which reacts with magnesium in ether and produces *sec*-butyl magnesium bromide *(B)*. It reacts with formaldehyde and produces 2-methyl-1-butanol. **(Grignard synthesis of alcohols)**

24. d. Esters react with two moles of Grignard reagents to produce tertiary alcohols. The ester that would react with methyl magnesium bromide to produce 2,3,3-trimethyl-2-pentanol is $CH_3CH_2C(CH_3)_2CO_2CH_3$. **(Grignard synthesis of alcohols)**

25. b. To accomplish this conversion the aldehyde should be treated with $NaBD_4$, converting it to $(CH_3CH_2CH_2CH_2CH_2CH(D)O)_4B$. Upon adding acid, this intermediate compound is converted to the product. **(Reduction of carbonyl compounds)**

26. d. All of these will give the desired product because acetic anhydride, acetyl chloride, and acetic acid will undergo an esterification reaction with cyclopentanol. **(Esterification)**

27. d. Tertiary alcohols do not oxidize under normal conditions because there is no H atom on the C atom with the OH. If a tertiary alcohol is strongly oxidized the molecule is cleaved and a complex mixture of products results. **(Oxidation of alcohols)**

28. b. $CH_3CH(OH)CH=CHCH_2OH$ is the product because $LiAlH_4$ reduces carbonyl groups to alcohols but cannot reduce double bonds. **(Reduction of carbonyl compounds)**

29. d. 2,4,6-Trinitrophenol is the strongest acid because the addition of three nitro groups, electron-withdrawing groups, helps to disperse the negative charge on the alkoxide ion. The pK_a of 2,4,6-trinitrophenol is 0.42, a strong organic acid. **(Phenol properties)**

30. d. 2,4,6-Tribromophenol is the principal product because the OH is a strong activating group and in the presence of a catalyst both *ortho* and *para* positions are brominated. The catalyst should be omitted to produce the *para* product only. **(Phenol reactions)**

31. c. 2,4-Di-*t*-butyl-5-methylphenol is the product that forms because the OH group is a strong *ortho-para* director. The position between the OH and CH₃ would be significantly less likely because of steric factors. **(Phenol reactions)**

32. a. *G* is 3,4-dimethyl-3-hexanol. *D* = 2-butanone, *E* = 2-bromobutane, *F* = *sec*-butyl magnesium bromide **(Alcohol reactions)**

33. d. Both PCC and the Collins reagents oxidize primary alcohols in aldehydes. Jones reagent oxidizes alcohols to acids. **(Oxidation of alcohols)**

34. c. Tertiary alcohols react almost instantly with the Lucas reagent. Primary alcohols react slowly under the same conditions. **(Alcohol reactions)**

35. e. When thionyl chloride reacts with this compound, the following allylic cation results which allows for the formation of both I and II. **(Alcohol reactions)**

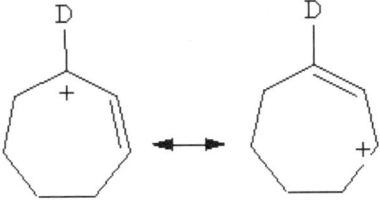

Grade Yourself

Circle the number of questions you missed, then fill in the total incorrect for each topic. If you answered more than three questions incorrectly, you need to focus on that topic. (If a topic has less than three questions and you had at least one wrong, we suggest you study that topic also. Read your textbook, a review book, or ask your teacher for help.)

Subject: Alcohols and Phenols

Topic	Question Numbers	Number Incorrect
Grignard synthesis of alcohols	1, 2, 4, 5, 21, 22, 23, 24	
Organolithium synthesis of alcohols	3	
Alcohol synthesis	6, 12, 15, 19	
Alcohol synthesis by carbonyl reduction	7, 20	
Diolsynthesis	8	
Alcohol condensation reactions	9	
Converting alcohols to alkyl halides	10	
Esterification	11, 26	
Alcohol reactions	13, 32, 34, 35	
Formation of sulfonate esters	14	
Phenol properties	16, 29	
Phenol reactions	17, 18, 30, 31	
Reduction of carbonyl compounds	25, 28	
Oxidation of alcohols	27, 33	

Ethers and Epoxides

14

Brief Yourself

Ethers are molecules with the general formula of R–O–R´. The R groups can be the same (R = R´), a symmetrical ether, or they can be different (R ≠ R´), an unsymmetrical ether. Ethers are composed of polar molecules because the R groups donate electron density into the more electronegative O atom; thus, most have nonzero dipole moments ($\mu > 0$). The boiling points of ethers are about the same as alkanes with similar molecular masses. The polar nature of ether molecules has little effect on their boiling points. Nonetheless they are polar solvents and can dissolve many polar molecules. Nonpolar molecules tend to be more soluble in ethers than in alcohols.

The common names of simple ethers are the names of the alkyl groups bonded to the O atom followed by the word *ether*. For example, CH_3–O–CH_2CH_3 is called methyl ethyl ether. Its IUPAC name is methoxyethane. For more complex ethers, the name of the alkoxy group, OR, is added to the name of the parent molecule. For example, consider 1-ethyl-4-ethoxycyclohexane.

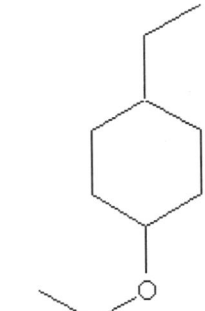

1-ethyl-4-ethoxycyclohexane

Cyclic ethers are heterocyclic compounds, those containing a noncarbon atom in a ring. Epoxides are three-membered cyclic ethers. The simplest epoxide is ethylene oxide, oxirane, or oxacyclopropane.

ethylene oxide
(oxirane, oxacyclopropane

Other common cyclic ethers include furan, tetrahydrofuran, pyran, and tetrahydropyran.

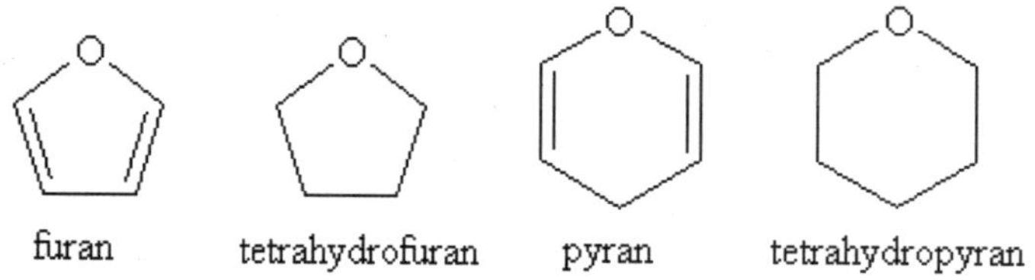

furan   tetrahydrofuran   pyran   tetrahydropyran

Some cyclic ethers can have more than one O atom. An example of such a molecule is 1,4-dioxane.

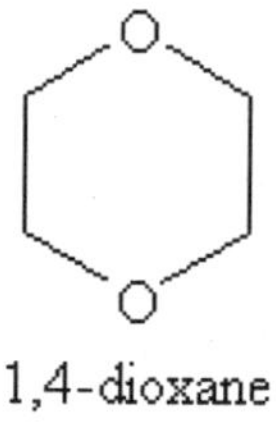

1,4-dioxane

Ethers can be synthesized using the Williamson ether synthesis. In this reaction, an alkoxide reacts with a primary alkyl halide.

$$R–O^- + R'–X \rightarrow R–O–R' + X^-$$

This reaction does not work for any compounds that readily undergo E2 elimination reactions, e.g., tertiary alkyl halides. Ethers can also be synthesized through acid-catalyzed condensation reactions. In this reaction, two alcohol molecules condense to produce a symmetrical ether when heated.

$$2R–OH \ (H_2SO_4) \rightarrow R–O–R + H_2O$$

One of the best ways to produce ethers is through solvomercuration-demercuration, also called alkoxymercuration-demercuration. Essentially, this is the same reaction as oxymercuration-demercuration used to synthesize alcohols but in place of water an alcohol is used. In the first step an alkene reacts with $Hg(OAc)_2$ and ROH, and in the second step the product is reduced with $NaBH_4$.

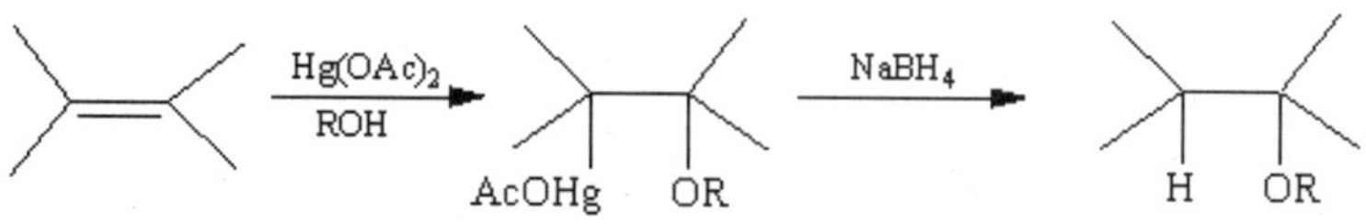

Epoxides are synthesized using peracids.

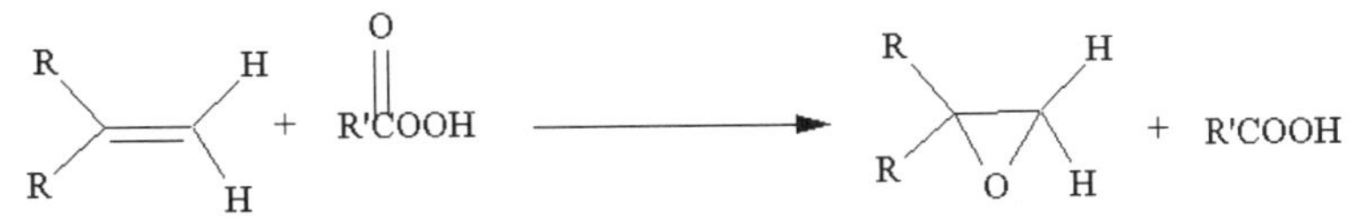

Another way to synthesize epoxides is to treat vicinal halohydrins with base.

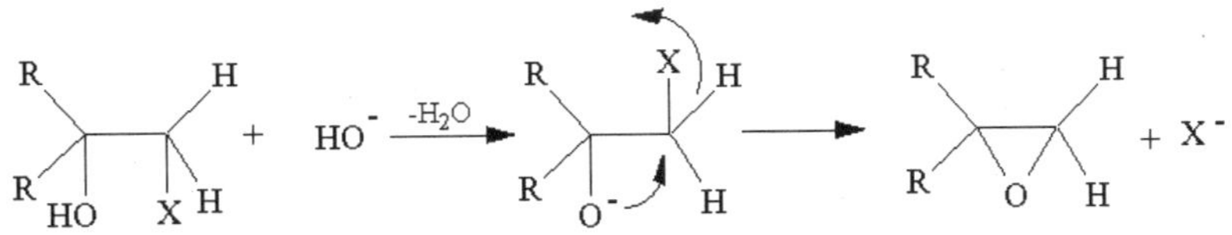

This reaction is essentially an intramolecular Williamson ether synthesis.

Ethers are not reactive when combined with most reagents. However, both HBr and HI, strong acids, cleave ether molecules when excess amounts of acid are used. The products of the reaction are two alkyl halides.

$$R–O–R' + HX(excess) \rightarrow RX + R'X$$

Epoxides are very reactive. Nucleophiles react with epoxides and open the ring. A good example of this reaction is the attack of Grignard reagents with epoxides.

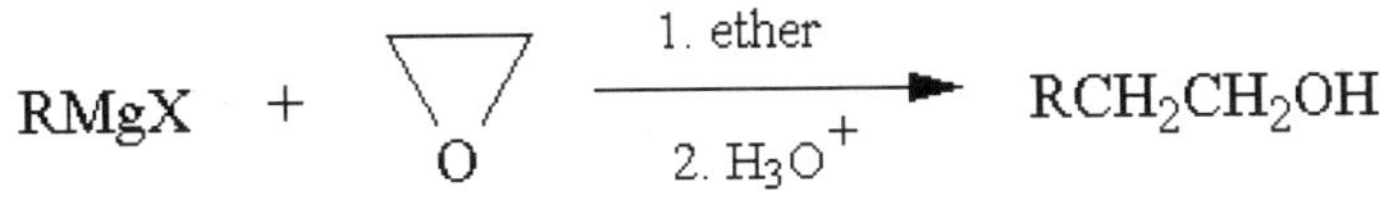

Epoxides also undergo acid-catalyzed ring-opening reactions. For example, ethylene oxide reacts with aqueous sulfuric acid and produces ethylene glycol. A similar reaction takes place if sulfuric acid and ethanol is mixed with ethylene oxide. The product of this reaction is 2-ethoxyethanol, $CH_3CH_2OCH_2CH_2OH$. If one mole of a strong acid such as HI reacts with ethylene oxide, 2-iodoethanol results. When excess HI is present, the product of this reaction is 1,2-diiodoethane.

Test Yourself

1. a. What is an ether? Give a specific example.

    b. What is an epoxide? Give a specific example.

2. Draw the structure of the cyclic ether tetrahydrofuran, also called oxolane.

3. a. Draw the structure of *bis*-(2-methoxyethyl) ether, also called diglyme.

    b. In what general class of ethers is diglyme?

4. Draw the structure of 1,2-epoxycyclopentane.

5. a. What type of intermolecular forces are found in diethyl ether?

    b. Is the normal boiling point of diethyl ether higher or lower than that of pentane?

6. Show that diethyl ether can form hydrogen bonds with water molecules.

7. a. What is a crown ether?
    b. Draw the structure of a crown ether.

8. Write the name for the following ether.

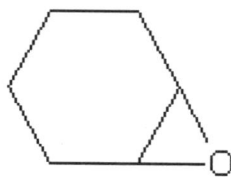

9. Draw the structure of 1,3,5-trioxane.

10. Starting with 1-propanol show how dipropyl ether can be prepared using acid-catalyzed condensation.

11. Show two ways in which benzyl *n*-butyl ether is synthesized using the Williamson ether synthesis.

12. Write an equation that shows how 2-ethoxy-3, 3-dimethylbutane can be synthesized using solvomercuration-demercuration.

13. What product(s) results when 1,2-dimethoxy-ethane is heated with excess HI?

14. Write a complete mechanism that shows the reaction of HI with dimethyl ether.

15. Draw the structure(s) of the product(s) of the reaction of *cis*-2-butene and peroxyacetic acid.

16. Write the equation for the reaction of *trans*-2-bromocyclopentanol with sodium hydroxide.

17. After phenylmagnesium bromide reacts with ethylene oxide in ether, the product is hydrolyzed and then heated in sulfuric acid. What is the product of the reaction?

18. Consider the following reaction.

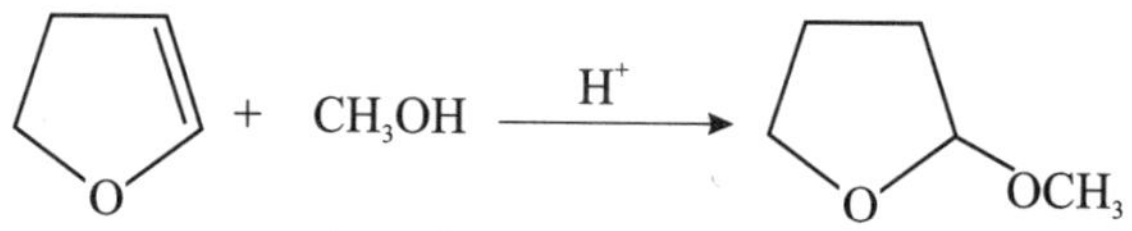

$$\text{(furan-like ring with O)} + CH_3OH \xrightarrow{H^+} \text{(tetrahydrofuran ring with OCH}_3\text{)}$$

Write a mechanism that explains this reaction.

19. Starting with cyclopentanol and necessary reagents, outline a synthesis of the following epoxide.

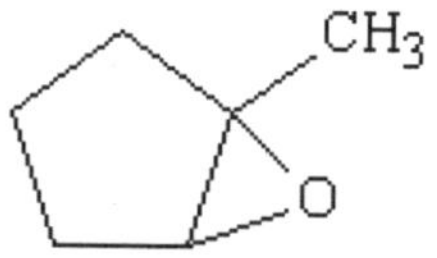

(cyclopentane ring fused with epoxide bearing CH₃)

20. Which of the following has the highest boiling point?

    a. pentane

    b. 1-butanol

    c. diethylether

    d. ethylene oxide

21. What is the principal product of the reaction of methyl bromide with the following compound?

    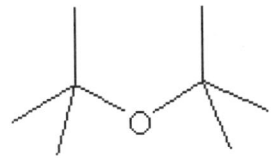

    a. dicyclohexyl ether

    b. dimethyl ether

    c. methyl cyclohexyl ether

    d. all of these

22. What is the principal product of the reaction of *t*-butyl bromide and sodium ethoxide?

    a. ethyl *t*-butyl ether

    b. diethyl ether

    c. di-*t*-butyl ether

    d. none of these

23. Which of the following sets of reagents gives the highest yield of the following ether?

    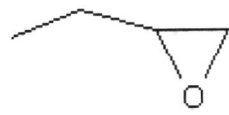

    a. Heating $(CH_3)_3COH$ with sulfuric acid

    b. Mixing $(CH_3)_3CO^- K^+$ and $(CH_3)_3CBr$

    c. First, mix 2-methylpropene with sulfuric acid and then allow the nucleophilic displacement with $(CH_3)_3CCl$

    d. First mix 2-methylpropene with $(CH_3)_3COH$ with $Hg(OCOCF_3)_2$ and then react the product with sodium borohydride and base

24. What is the product of the reaction of methyl ethyl ether with $LiAlH_4$?

    a. CH_3OH and CH_3CH_2OH

    b. CH_3OLi and CH_3CH_2OLi

    c. $H_2C=O + CH_3CH_2OH$

    d. no reaction

25. What product(s) result when the following ether is treated with excess HBr?

    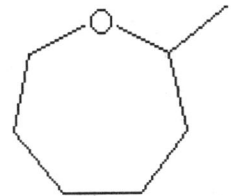

    a. $HO–CH_2CH_2CH_2CH_2CH_2CH_2CH_2-Br$

    b. $Br–CH_2CH_2CH_2CH_2CH_2CH_2CH_2-Br$

    c. $Br–CH(CH_3)CH_2CH_2CH_2CH_2CH_2-Br$

    d. $Br–CH(CH_3)CH_2CH_2CH_2CH_2CH_2-OH$

    e. none of these

26. What is the principal product that results when the following epoxide reacts with ethanol and sodium ethoxide?

    a. $CH_3CH_2CH(OH)CH_2OCH_2CH_3$

    b. $CH_3CH_2CH(OCH_3)CH_2OCH_2CH_3$

    c. $CH_3CH_2CH(OCH_2CH_3)CH_2OCH_2CH_3$

    d. $CH_3CH_2CH(OH)CH_2OH$

27. What is the principal product that results when the following epoxide reacts with sodium methoxide and methanol?

   a. *cis*-2-methoxycyclohexanol

   b. *trans*-2-methoxycyclohexanol

   c. 1-methoxycyclohexanol

   d. none of these

28. Cyclopentanol is dehydrated with sulfuric acid, producing Compound W. W reacts with peroxyacetic acid, producing Compound X. X is treated with one mole of HBr, producing Compound Y. What is the identity of Y?

   a. *trans*-1,2-dibromocyclopentane

   b. *cis*-1,2-dibromocyclopentane

   c. *trans*-1,2-cyclopentanediol

   d. *cis*-1,2-cyclopentanediol

   e. none of these

29. What is the product of methyloxirane with phenyl magnesium bromide followed by acid hydrolysis?

   a. C_6H_5–$CH_2CH(OH)CH_3$

   b. C_6H_5–$CH(OH)CH_2CH_3$

   c. C_6H_5–$CH_2CH_2CH_2$–OH

   d. none of these

30. Which of the following is a correct statement concerning the spectroscopy of ethers?

   a. Ethers usually exhibit a moderate to strong C–O single bond stretch peak around 1000 to 1200 cm^{-1}.

   b. The mass spectra of ethers typically shows a peak where the molecule is cleaved alpha to the O atom.

   c. The ^{13}C NMR spectrum of an ether shows peaks in the range of 50 to 80 ppm (δ) that correspond to the C atoms bonded to the O atom.

   d. all of the above

   e. none of the above

✔ Check Yourself

1. a. An ether is a compound that has the general formula of R–O–R′. An example of an ether is diethyl ether, CH_3CH_2–O–CH_2CH_3.
   b. An epoxide is a cyclic ether in which the O atom is a component of a three-membered ring. The simplest epoxide is ethylene oxide, also called oxirane. (**Ethers and epoxides**)

   ethylene oxide (oxirane, oxacyclopropane)

2. The structure of tetrahydrofuran, THF, is as follows:

   THF
   (**Cyclic ethers**)

3. a. The structure of diglyme is as follows.

    $CH_3-O-CH_2CH_2-O-CH_2CH_2-O-CH_3$

    b. Diglyme is a polyether, one with more than one ether linkage. **(Ether nomenclature)**

4. The structure of 1,2-epoxycyclopentane is as follows. **(Epoxide structures)**

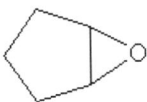

5. a. Diethyl ether, and ethers in general, are polar molecules which means their intermolecular forces are both dipole-dipole interactions and London forces.

    b. Diethyl ether and pentane have similar boiling points because the polar nature of diethyl ether molecules has little effect on its boiling point. The normal boiling points of diethyl ether and pentane are $34.6^\circ C$ and $36^\circ C$, respectively. **(Physical properties of ethers)**

6. Diethyl ether can accept a hydrogen bond from water as follows.

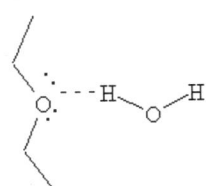

    (Physical properties of ethers)

7. a. A crown ether is a cyclic ether with four or more ether linkages in a ring of 12 or more atoms.

    b. An example of a crown ether is 18-crown-6.

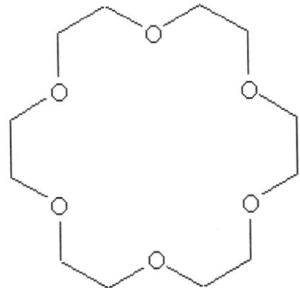

    (Crown ethers)

8. The name of this ether is 1,2-epoxycyclohexane. **(Epoxide nomenclature)**

9. The structure of 1,3,5-trioxane is as follows. **(Cyclic ethers)**

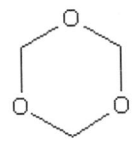

    1,3,5-trioxane

10. In the acid-catalyzed condensation reaction, two alcohol molecules join producing an ether and water. The following shows the reaction of two moles of 1-propanol to produce dipropyl ether and water. This reaction takes place when heated with H_2SO_4. (**Synthesis of ethers**)

$$CH_3CH_2CH_2OH + CH_3CH_2CH_2OH \rightarrow CH_3CH_2CH_2–O–CH_2CH_2CH_3 + H_2O$$

11. In the Williamson synthesis, an alkoxide does an S_N2 displacement of a halide in an alkyl halide. Thus, one way to produce benzyl n-butyl ether is as follows.

$$CH_3CH_2CH_2CH_2O^- + C_6H_5–CH_2Br \rightarrow CH_3CH_2CH_2CH_2\text{-}O\text{-}CH_2C_6H_5 + H_2O$$

    A second way to produce the same product is as follows.

$$CH_3CH_2CH_2CH_2Br + C_6H_5–CH_2O^- \rightarrow CH_3CH_2CH_2CH_2–O–CH_2C_6H_5 + H_2O$$

    (**Synthesis of ethers**)

12. 2-Ethoxy-3,3-dimethylbutane can be synthesized using solvomercuration-demercuration by reacting 3,3-dimethyl-1-butene with $Hg(OAc)_2$ and CH_3CH_2OH followed by $NaBH_4$ and OH^-.

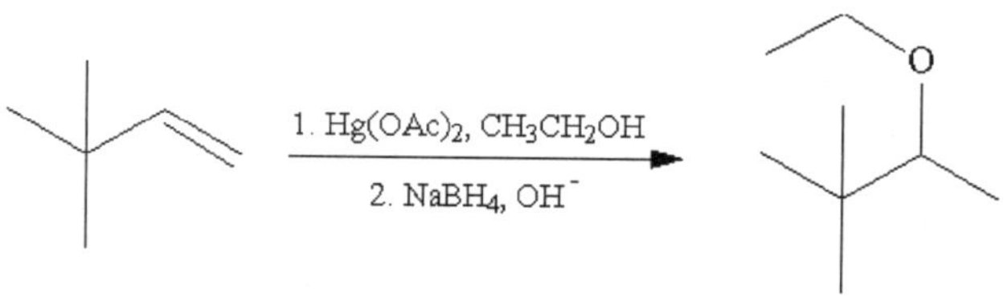

    3,3-dimethyl-1butene   2-ethoxy-3,3-dimethylbutane

    (**Synthesis of ethers**)

13. 1,2-Dimethoxyethane, $CH_3–O–CH_2CH_2–O–CH_3$ is cleaved when heated with HI, yielding two molecules of CH_3I and one molecule of ICH_2CH_2I. (**Ether reactions**)

14. The mechanism for the acid cleavage by HI of dimethyl ether is as follows.

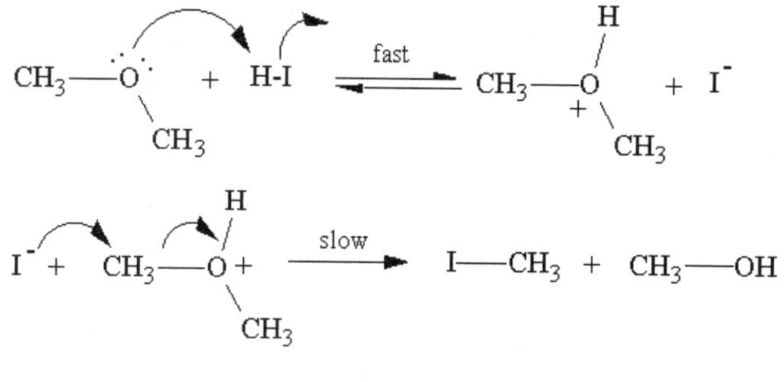

    The resulting methanol is protonated by the HI forming $CH_3OH_2^+$ and I^- does an S_N2 displacement, yielding another methyl iodide molecule. (**Ether reactions**)

15. The products of the reaction are cis-2,3-dimethyloxirane and acetic acid, CH_3COOH. The structure of cis-2,3-dimethyloxirane is as follows.

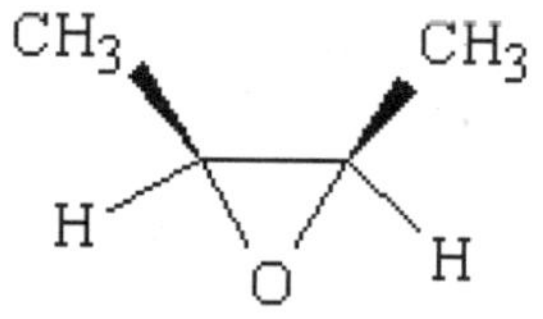

    cis-2,3-dimethyloxirane

    (**Epoxide synthesis**)

16. The product of the reaction is 1,2-epoxycyclopentane.

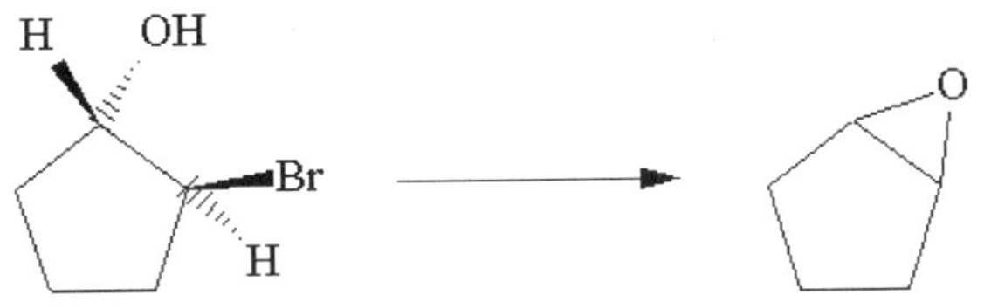

trans-2-bromocyclopentanol   1,2-epoxycyclopentane

(Epoxide synthesis)

17. Phenylmagnesium bromide, C_6H_5–MgBr, reacts with ethylene oxide, and after hydrolysis forms 2-phenylethanol. Heating 2-phenylethanol with sulfuric acid initiates a condensation reaction that produces the following ether.

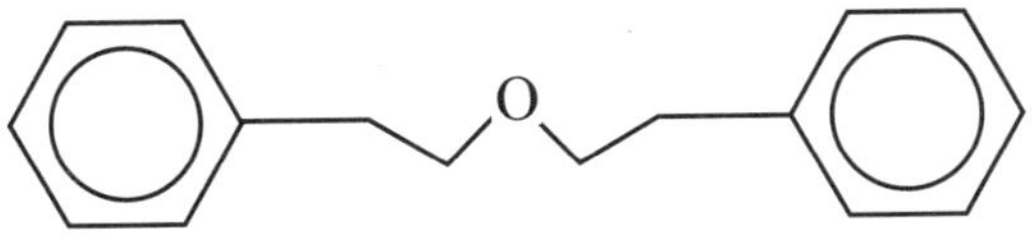

(Ether/epoxide reactions)

18. The first step is the protonation of the ring by the acid.

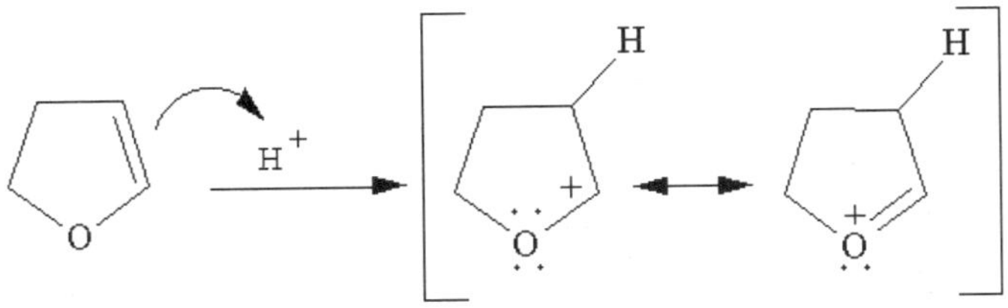

A resonance-stabilized carbocation forms because the O atom in the ring can donate electron density to the carbocation C atom. In the final step, the O atom of the methanol bonds to the carbocation and then loses its proton to the solvent.

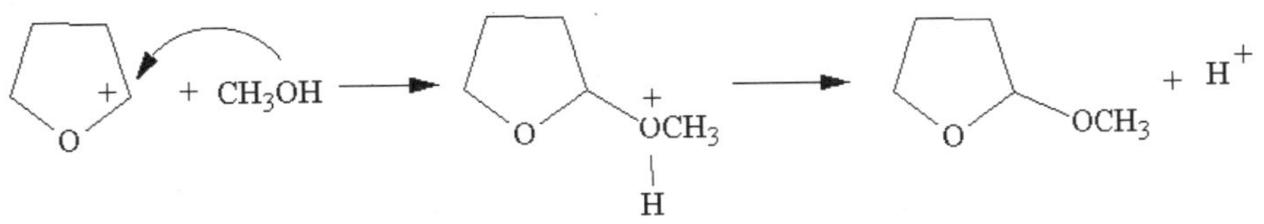

(Ether reactions)

19. Start by using chromic acid to oxidize cyclopentanol to cyclopentanone. React cyclopentanone with methyl magnesium bromide, CH_3MgBr, in ether followed by hydrolysis to produce 1-methylcyclopentanol. This alcohol is then dehydrated with sulfuric acid to 1-methylcyclopentene. React the 1-methylcyclopentene with peroxyacetic acid to produce the desired product. **(Epoxide synthesis)**

20. b. 1-Butanol has the highest boiling point. Alcohols with similar molecular masses have higher boiling points than ethers or alkanes because of their hydrogen bonding. **(Physical properties of ethers)**

21. c. Methyl cyclohexyl ether is the only product that results from this Williamson ether synthesis. In this reaction, the alkoxide, $C_6H_{11}O^-$, substitutes for the Br atom in methyl bromide. **(Synthesis of ethers)**

22. d. Tertiary alkyl halides such as *t*-butyl chloride undergo E2 elimination reactions with alkoxides and produce alkenes. Thus, the product of this reaction is not an ether but isobutene. **(Synthesis of ethers)**

23. d. 2-Methylpropene reacts with $(CH_3)_3COH$ and $Hg(OCOCF_3)_2$ followed by sodium borohydride and base in solvomercuration-demercuration. **(Synthesis of ethers)**

24. d. Ethers do not undergo reduction with $LiAlH_4$. **(Ether reactions)**

25. c. The product of this reaction is $Br-CH(CH_3)CH_2CH_2CH_2CH_2CH_2-Br$. Strong acids such as HBr cleave ether molecules and produce alkyl halides. **(Ether reactions)**

26. a. $CH_3CH_2CH(OH)CH_2OCH_2CH_3$ is the product of this reaction. In this reaction, the nucleophilic ethoxide ion, $CH_3CH_2O^-$, attacks the epoxide at the least hindered position. The resulting new alkoxide takes in a proton, and an OH group forms on the other C atom. **(Epoxide reactions)**

27. b. *trans*-2-Methoxycyclohexanol is the product of this reaction because the attack of the nucleophile, CH_3O^-, is anti to the epoxide O atom. Thus, the methoxy group must be *trans* to the OH group. **(Epoxide reactions)**

28. e. None of these. The product of the reaction is *trans*-2-bromocyclopentanol (*Y*). Cyclopentanol dehydrates and forms cyclopentene (*W*). It reacts with peroxyacetic acid and produces 1,2-epoxycyclopentane (*X*). Treatment with HBr produces *trans*-2-bromocyclopentanol. **(Epoxide reactions)**

29. a. $C_6H_5-CH_2CH(OH)CH_3$ is the product of the reaction. The phenyl magnesium bromide attacks the least hindered position of the epoxide, producing $C_6H_5-CH_2CH(OMgBr)CH_3$, which upon hydrolysis gives $C_6H_5-CH_2CH(OH)CH_3$. **(Epoxide reactions)**

30. d. All of these statements are true about the spectra of ethers. **(Spectroscopic analysis of ethers)**

Grade Yourself

Circle the number of questions you missed, then fill in the total incorrect for each topic. If you answered more than three questions incorrectly, you need to focus on that topic. (If a topic has less than three questions and you had at least one wrong, we suggest you study that topic also. Read your textbook, a review book, or ask your teacher for help.)

Subject: Ethers and Epoxides

Topic	Question Numbers	Number Incorrect
Ethers and epoxides	1	
Cyclic ethers	2, 9	
Ether nomenclature	3	
Epoxide structures	4	
Physical properties of ethers	5, 6, 20	
Crown ethers	7	
Epoxide nomenclature	8	
Synthesis of ethers	10, 11, 12, 21, 22, 23	
Ether reactions	13, 14, 17, 18, 24, 25	
Epoxide synthesis	15, 16, 19	
Epoxide reactions	17, 26, 27, 28, 29	
Spectroscopic analysis of ethers	30	

Aldehydes and Ketones

15

Brief Yourself

Aldehydes are compounds with the general formula of RCHO, and ketones have the general formula of RCOR′. Both aldehydes and ketones have a carbonyl group, C=O. The properties of both classes of compounds depend on the properties of the carbonyl group.

A carbonyl group is similar to a C–C double bond except that it is shorter and stronger. It is also polar because of the more electronegative O atom. The carbonyl C atom carries a partial positive charge and the carbonyl O atom carries a partial negative charge.

The IUPAC names of ketones are derived from the longest continuous chains with the carbonyl group. The ending for ketones is *one*. Some examples of ketones are as follows.

2-methyl-3-octanone   cyclopentanone

The IUPAC names for aldehydes are derived in a similar manner to ketones but the ending *al* is added to the parent alkane. No number is needed to designate the location of the aldehyde group because it is always at the end of a chain. When it is a substituent group bonded to another structure, the ending *carbaldehyde* is used. Examples of aldehyde names are as follows.

2,2-dichloro-3-methylhexanal   2-bromocycloheptanecarbaldehyde

The polar carbonyl group creates greater dipole-dipole forces than found in either alkanes or ethers; thus, carbonyl compounds have higher boiling points than alkanes or ethers with similar molecular masses. Because aldehydes and ketones do not have a H atom bonded to an O atom, they do not donate H bonds. Nonetheless they can accept H bonds from alcohols and amines, compounds with O–H and N–H bonds; hence, they are good solvents for such compounds.

Aldehydes and ketones can be synthesized in many ways. They can be formed from the oxidation of alcohols. Ketones result from the oxidation of secondary alcohols. Aldehydes are produced when primary alcohols are oxidized with Collins reagent or PCC. Another method to produce aldehydes and ketones is through the ozonolysis of alkenes. The alkene molecule is cleaved at the double bond and usually two carbonyl compounds result. Aromatic ketones can be produced by the Friedel-Crafts acylation reaction, in which an acyl chloride bonds to an aromatic ring in the presence of a strong Lewis acid catalyst. Aldehydes and ketones can be produced from alkynes. If an alkyne is treated with aqueous H_2SO_4 and $HgSO_4$, a ketone results.

An excellent method of producing aldehydes and ketones is through the alkylation of 1,3-dithianes. In this reaction, 1,3-dithiane is first treated with butyl lithium followed by a primary alkyl halide. If only one R group is bonded to the 2-position of 1,3-dithiane, hydrolysis with acid and mercury(II) chloride produces an aldehyde. If two R groups are bonded to the 2-position of 1,3-dithiane, hydrolysis with acid and mercury(II) chloride produces a ketone.

Ketones can be synthesized from carboxylic acids. An acid is first treated with two moles of an organolithium compound, RLi, which is then hydrolyzed to a ketone.

$$RCOOH + 2R'Li \rightarrow R\text{–}C(OLi)_2\text{–}R' \rightarrow RCOR'$$

Ketones are also synthesized from nitriles, RCN. A nitrile is treated with a Grignard reagent or organolithium compound and an imine magnesium salt results. This compound is hydrolyzed to a ketone.

$$R\text{–}CN + R'\text{–}MgX \rightarrow RR'C=N\text{–}MgX + H_3O^+ \rightarrow RCOR'$$

Aldehydes and ketones are reactive compounds. The principal type of reaction that they undergo is nucleophilic addition. In this reaction, a nucleophile, Nu:$^-$, attacks the carbonyl C atom, and the group initially bonded to the nucleophile, H$^+$ (for example), then bonds to the carbonyl O atom.

$$H-Nu + C=O \rightarrow Nu-C-O-H$$

An example of a nucleophilic addition to a carbonyl group is the addition of a Grignard reagent, RMgX, which produces alcohols after hydrolysis.

$$RMgX + R'CHO \rightarrow RCH(OH)R'$$

Aldehydes and ketones can also be reduced with $NaBH_4$ or $LiAlH_4$, sources of nucleophilic H^-.

Some aldehydes and ketones react with water in nucleophilic addition reactions and produce hydrates, geminal diols.

$$R_2C=O + H_2O \rightleftarrows R_2C(OH)_2$$

Aldehydes more readily hydrate than ketones because the R groups of ketones stabilize the carbonyl C atom. When alcohols, R'OH, react with aldehydes, they produce hemiacetals (RCH(OH)(OR')) and acetals (RCH(OR')$_2$), and when they react with ketones, they produce hemiketals ($R_2C(OH)(OR')$) and ketals ($R_2C(OR')_2$). In the IUPAC system, both hemiketals and ketals are also referred to as hemiacetals and acetals.

$$RCHO + R'OH \rightarrow RCH(OH)(OR') + R'OH \rightarrow RCH(OR')_2$$

$$R_2C=O + R'OH \rightarrow R_2C(OH)(OR') + R'OH \rightarrow R_2C(OR')_2$$

Aldehydes and ketones also react in a nucleophilic addition reaction with HCN to produce cyanohydrins, RCH(OH)CN.

$$RCHO + HCN \rightarrow RCH(OH)CN$$

Aldehydes and ketones undergo nucleophilic addition-condensation reactions with ammonia and primary amines to produce imines, –C=NH–R. The intermediate compound that results after the nucleophilic addition is a carbinolamine.

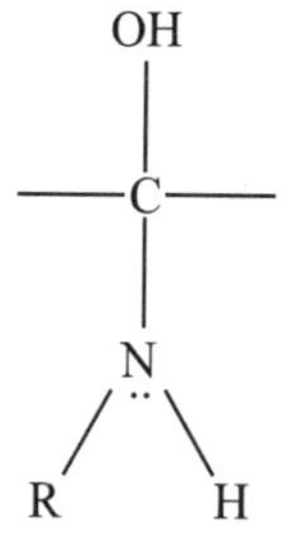

carbinolamine

It immediately dehydrates to an imine, RCH=N–R´.

$$RCHO + R´–NH_2 \rightarrow RCH(OH)(NHR´) \rightarrow RCH=N–R´$$

Besides ammonia and primary amines, aldehydes and ketones undergo similar reactions with many nitrogen derivatives. Hydroxylamine, HO–NH$_2$, reacts with carbonyl compounds and forms oximes, R$_2$C=N–OH. Hydrazine, H$_2$N–NH$_2$, reacts in a similar manner and forms hydrazones, R$_2$C=N–NH$_2$. Phenylhydrazine, C$_6$H$_5$–NH–NH$_2$, reacts and forms phenyl hydrazones, R$_2$C=N–NH–C$_6$H$_5$. Finally, semicarbazide, H$_2$NCONH–NH$_2$, reacts and forms semicarbazones, R$_2$C=N–NHCONH$_2$.

Secondary amines react with aldehydes and ketones and produce enamines.

enamine

This reaction goes through a carbinolamine intermediate.

$$–CHO + R_2´NH \rightarrow –CH(OH)(NR_2´) \rightarrow –CH=CH–NR_2´$$

Aldehydes can be oxidized by a variety of oxidizing agents to carboxylic acids, RCOOH. Some common oxidizing agents that may be used are chromic acid, permanganate, and Ag$^+$. A common laboratory test for aldehydes is the Tollens silver mirror test. In this reaction, an aldehyde reacts with Ag(NH$_3$)$_2^+$ in base and produces metallic Ag that mirrors the inside of the beaker or test tube where the reaction takes place.

The carbonyl group in aldehydes and ketones can be reduced to methylene, CH$_2$, using either the Clemmensen reduction (Zn(Hg), HCl) or the Wolff-Kishner reduction (H$_2$NNH$_2$ followed by KOH and heat).

Test Yourself

1. Draw the structure of 4-bromo-2-heptanone.

2. Draw the structure of 2,3-diphenylbutanedial.

3. Write both the common and IUPAC names for the following molecule.

4. Write the IUPAC name for the following molecule.

5. Write a paragraph that explains the structure and properties of a carbonyl group. Is the carbonyl group polar or nonpolar? Explain.

6. List two ways that aldehydes and ketones can be produced from hydrocarbons. Give a specific example of each.

7. How could the following alcohol be converted to cyclopentanecarbaldehyde?

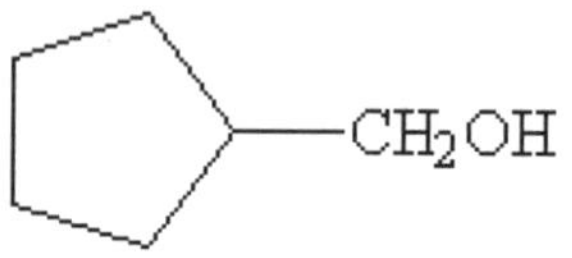

8. Write an equation that shows the synthesis of 1-phenyl-1-butanone using an organocadmium reagent.

9. What two reagents could be used to reduce propanal to propane?

10. Write the mechanism for the acid-catalyzed formation of the hemiacetal produced from ethanol and acetaldehyde.

11. Draw the structure of the cyclic acetal that forms when cyclohexanone reacts with 1,3-propanediol.

12. a. What is the product of the reaction of acetone with aqueous NaCN followed by sulfuric acid?

    b. What class of compound results from this reaction?

13. a. Draw the structure of the product of the reaction of benzophenone, $C_6H_5COC_6H_5$, and cyclopentylamine?

    b. What class of compound results from this reaction?

14. a. Draw the structure of the product of the reaction of 4-heptanone with pyrrolidine.

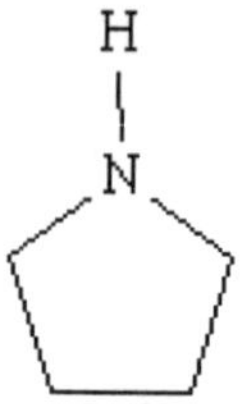

pyrrolidine

    b. What class of compound results from this reaction?

15. Draw the structure of the product of the reaction of acetophenone and semicarbazide.

16. What two reactants are required to produce the following alkene using a Wittig reaction?

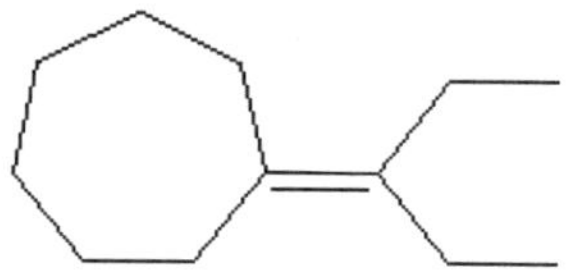

17. What two products can result when the following ketone is reduced with $NaBH_4$?

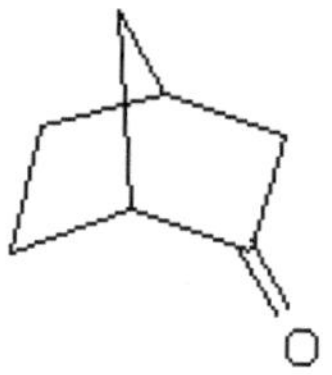

18. Using only acetyl chloride, CH_3COCl, and necessary inorganic reagents show how acetal, $CH_3CH(OCH_2CH_3)_2$, can be synthesized.

19. Show a synthesis of the following compound using 1,3-dithiane.

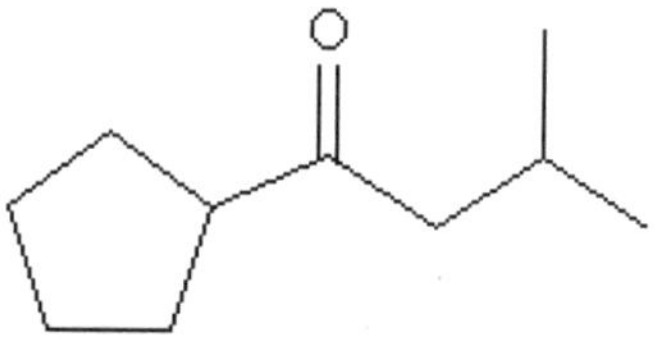

20. Which of the following reagents could be used to convert acetyl chloride to acetaldehyde?

    a. H_2, Pd/BaSO$_4$, heat

    b. LiAlH$_4$

    c. H_2, Pd

    d. none of these

21. Which of the following hydrates to the greatest extent?

    a. formaldehyde

    b. acetaldehyde

    c. propionaldehyde

    d. acetone

22. What type of compound is the following?

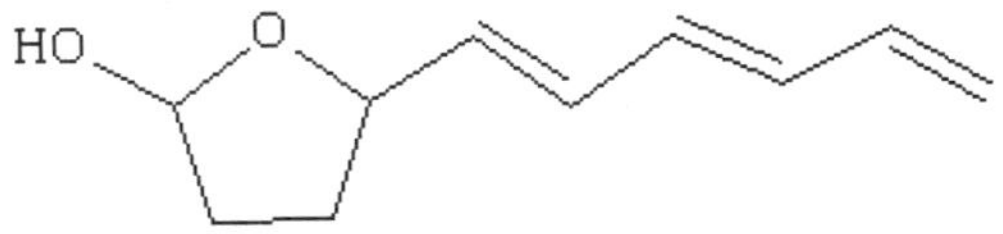

    a. acetal

    b. hemiacetal

    c. ketal

    d. hemiketal

    e. none of these

23. Which of the following will **not** react with hexanal to produce an imine?

    a. $(CH_3)_3CNH_2$

    b. $C_6H_5–NH_2$

    c. $(CH_3)_2NH$

    d. all of these produce imines

24. Which of the following is the product of the reaction of phenylhydrazine and benzaldehyde?

    a. $C_6H_5–CH=NNH–C_6H_5$

    b. $C_6H_5–CH=NNH_2$

    c. $C_6H_5–CH=NN(C_6H_5)_2$

    d. $C_6H_5–C(CH_3)=NNH–C_6H_5$

    e. none of these

25. What is the reaction intermediate that forms in the enamine formation reaction?

    a. oxime

    b. ylide

    c. semicarbazide

    d. carbinolamine

    e. none of these

26. Which of the following gives a positive Tollens test?

    a. benzaldehyde

    b. pentanal

    c. 2-methylpropanedial

    d. none of these

    e. all of these

27. Which of the following reacts most rapidly with HCN?

    a. CH_3COCH_3

    b. CH_3CHO

    c. $Cl–CH_2CHO$

    d. CCl_3CHO

28. Which of the following pairs of compounds could be used to prepare 2-methyl-1-phenyl-1-propene using the Wittig method?

    a. benzaldehyde, *n*-propyl bromide

    b. benzaldehyde, isopropyl bromide

    c. benzyl bromide, propanal

    d. phenyl bromide, acetone

    e. none of these can be used

    f. all of these can be used

29. Which of the following could be used to produce 1,3-dioxolane?

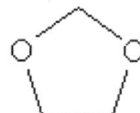

1,3-dioxolane

a. $HOCH_2CH_2OH$ and formaldehyde in acid

b. $HOCH_2CH_2CH_2OH$ and formaldehyde in acid

c. $HOCH_2CH_2CH_2OH$ and acetone in acid

d. $HOCH_2CH_2OH$ and acetone in acid

e. a and b

f. c and d

30. A compound with the formula $C_6H_{12}O$ gives a negative Tollens test and a positive iodoform test. It produces a semicarbazone and can be reduced to hexane. What is the identity of the compound?

a. $CH_3CH_2CH_2CH_2CH_2CHO$

b. $CH_3CH_2CH_2CH_2COCH_3$

c. $CH_3CH_2CH_2COCH_2CH_3$

d. $CH_3CH(CH_3)COCH_2CH_3$

e. none of these

31. Which of the following is **not** correct about the spectra of aldehydes and ketones?

a. In the IR spectrum, most show a strong stretching vibration near 1710 cm^{-1}.

b. In the ^1H NMR spectrum, the chemical shift of the aldehydic proton, –CHO, is in the range of 5 to 6 ppm (δ).

c. In the ^{13}C NMR spectrum, the chemical shift of the carbonyl C atom is near 200 ppm (δ).

d. In the mass spectrum, many produce acylium ions, $R-C\equiv O^+$.

e. all of the above

f. none of the above

32. Consider the following four compounds.

I   II   III   IV

Which of the following correctly ranks these carbonyl compounds from lowest to highest amount of hydrate present at equilibrium when reacted with water?

a. $I < II < III < IV$

b. $IV < III < II < I$

c. $II < I < III < IV$

d. $III < IV < II < I$

e. none of these

33. What product results when 2-pentanone is first treated with $NaBH_4$ followed by hydrolysis with D_2O?

a. $CH_3CH(OH)CH_2CH_2CH_3$

b. $CH_3CD(OH)CH_2CH_2CH_3$

c. $CH_3CH(OD)CH_2CH_2CH_3$

d. $CH_3CD(OD)CH_2CH_2CH_3$

e. none of these

34. Which of the following pathways produces 2-hexanone?

a. 1-hexyne is treated with H_2SO_4, $HgSO_4$, and H_2O

b. 3-methyl-2-heptene is treated with O_3 followed by hydrolysis

c. *n*-butyl magnesium bromide reacts with acetaldehyde, followed by hydrolysis, and then chromic acid oxidation

d. all of these

35. Which of the following will convert octanoyl chloride, $CH_3CH_2CH_2CH_2CH_2CH_2CH_2COCl$, to decane?

a. 1. $LiAlH_4$, 2. CH_3CH_2Cl, H^+, 3. heat

b. 1. $LiCu(CH_2CH_3)_2$, 2. $KMnO_4$

c. 1. 2 CH_3CH_2OH, H^+, 2. Zn(Hg) and HCl

d. 1. $(CH_3CH_2)_2Cd$, 2. H_2NNH_2, base, and heat

Check Yourself

1. The structure of 4-bromo-2-heptanone is as follows.

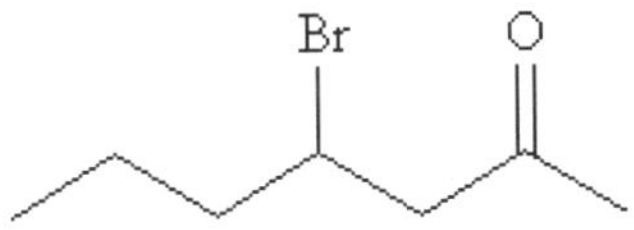

4-bromo-2-heptanone

   (Ketone nomenclature)

2. The structure of 2,3-diphenylbutanedial is as follows.

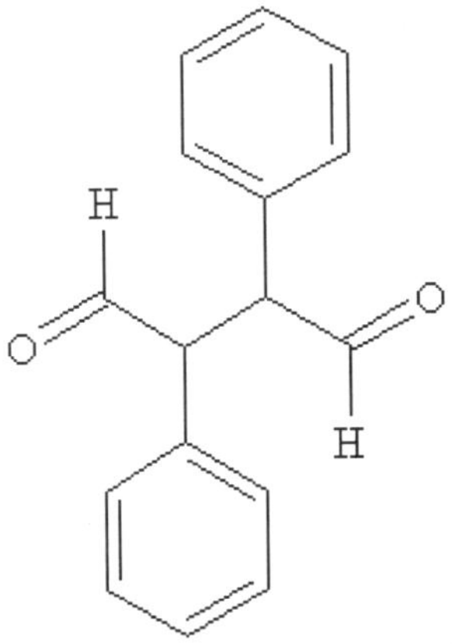

2,3-diphenylbutanedial

   (Aldehyde nomenclature)

3. The IUPAC name for this molecule is 1-hepten-4-one. Its common name is allyl *n*-propyl ketone. **(Ketone nomenclature)**

4. The IUPAC name for this molecule is 2-methylcyclohexanecarbaldehyde. **(Aldehyde nomenclature)**

5. A carbonyl group is a C–O double bond. To form the π bond of the double bond, both the C and O atoms are sp^2 hybridized. The geometry of the carbonyl group is trigonal planar, with bond angles close to 120°. The carbonyl group is polar because the O atom is more electronegative than the C atom. The following shows the two resonance structures that best represent a carbonyl group.

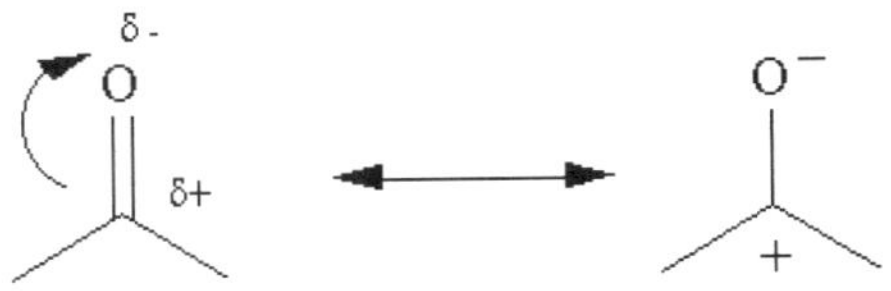

   (Carbonyl structure)

6. One way is the ozonolysis of alkenes. An example is the ozonolysis of 1-butene. It is first treated with ozone and the product is hydrolyzed with Zn and water.

$$H_2C=CHCH_2CH_3 + O_3 \rightarrow H_2C=O + CH_3CH_2CHO$$

The other way is to hydrate alkynes. An example is the hydration of 1-butyne using $HgSO_4$ and H_2SO_4 to 2-butanone.

$$HC\equiv CCH_2CH_3 + H_2O \rightarrow CH_3COCH_2CH_3$$

(Carbonyl compound synthesis)

7. Alcohols can be oxidized to aldehydes using the Collins reagent, which is a pyridine-chromium trioxide complex, $2(C_5H_5N)\cdot CrO_3$ in methylene chloride.

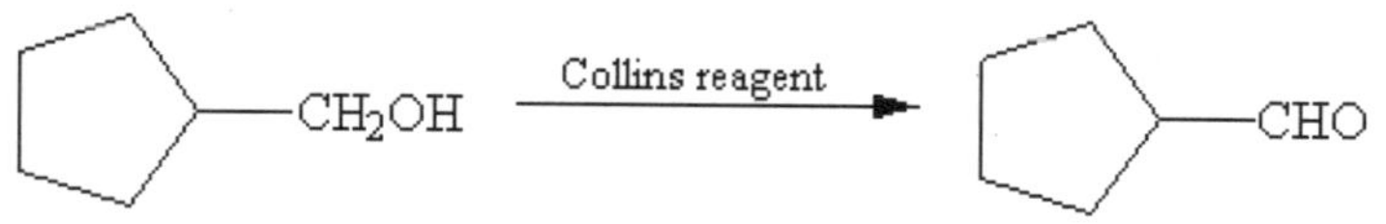

(Aldehyde synthesis)

8. Di-*n*-propylcadmiun $(CH_3CH_2CH_2)_2Cd$, which is prepared from *n*-propyl magnesium bromide and $CdCl_2$ in ether, reacts with benzoyl chloride to produce 1-phenyl-1-butanone. **(Ketone synthesis)**

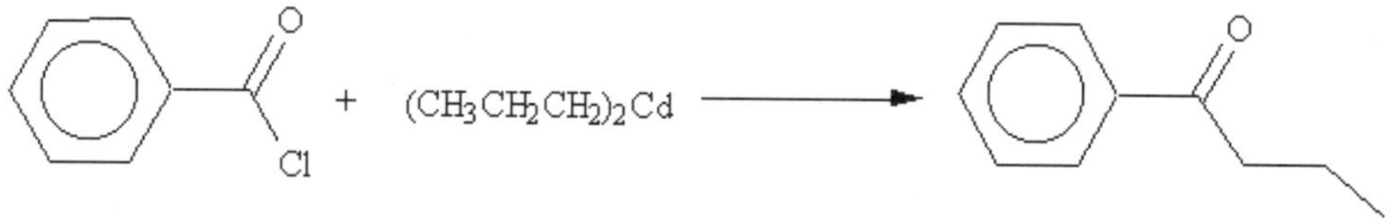

9. Aldehydes can be reduced to alkanes using either the Clemmensen reduction or the Wolff-Kishner reduction. In the Clemmensen reduction, a Zn amalgam and hydrochloric acid are used. In the Wolff-Kishner reduction, the aldehyde is heated with potassium hydroxide and hydrazine in a high-boiling alcohol. **(Aldehyde reactions)**

10. The mechanism for the formation of the hemiacetal of acetaldehyde and ethanol is as follows.

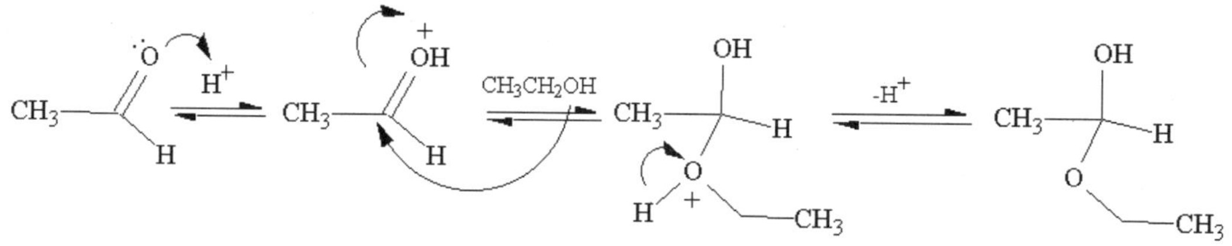

(Hemiacetal formation)

11. The cyclic acetal between cyclohexanone and 1,3-propanediol is as follows.

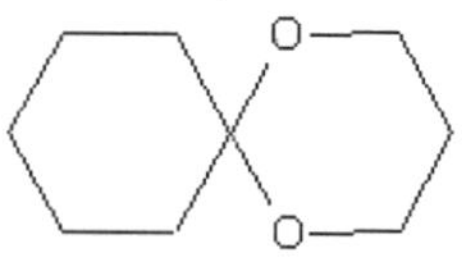

(Acetal formation)

12. a. The product of this reaction is as follows.

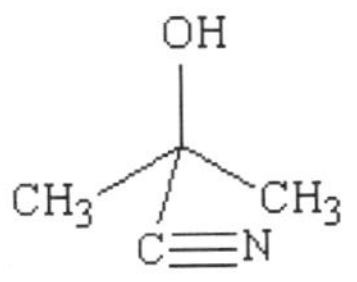

b. The product of this reaction is a cyanohydrin, specifically acetone cyanohydrin. (**Cyanohydrin formation**)

13. a. The product of this reactions is as follows.

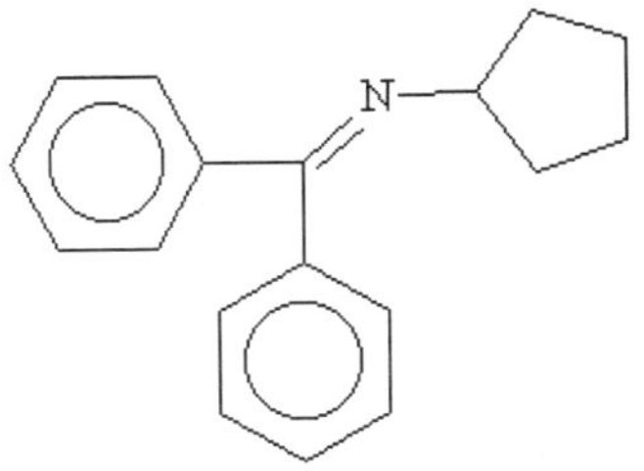

b. The product of this reaction is an imine. (**Imine formation**)

14. a. The product of the reaction is as follows.

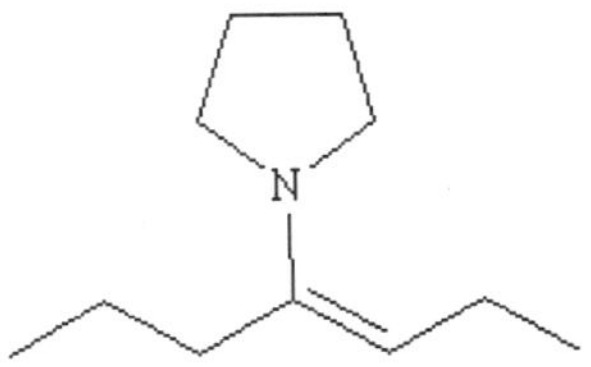

b. The product of this reaction is an enamine. (**Enamine formation**)

15. The product of the reaction of acetophenone and semicarbazide is acetophenone semicarbazone.

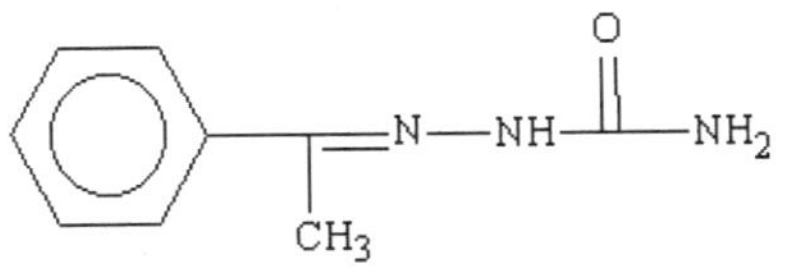

acetophenone semicarbazone

(**Semicarbazone formation**)

16. The two reactants needed to produce the desired alkene is as follows.

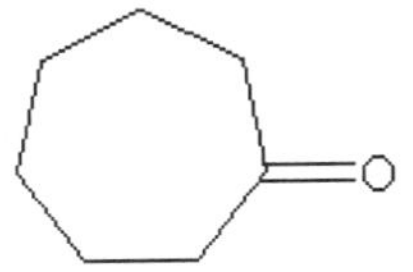

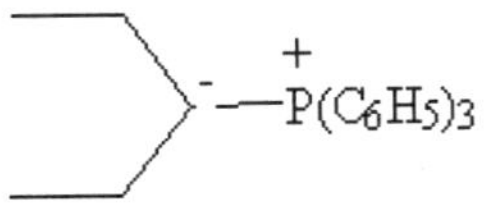

This reaction could also use 3-pentanone and cycloheptylidenetriphenylphosphorane. (**Wittig reaction**)

17. The two products of the reactions are as follows.

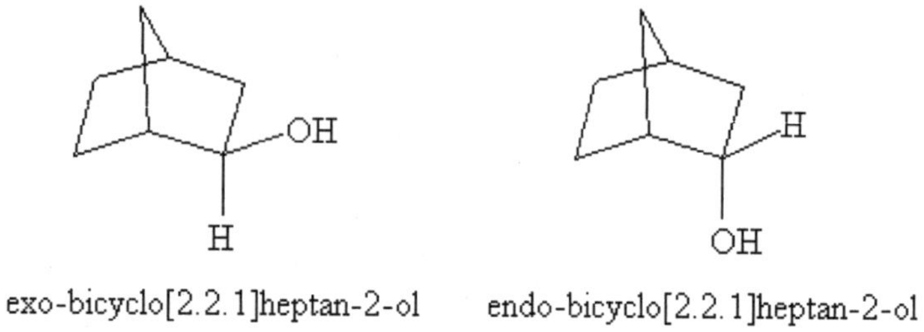

exo-bicyclo[2.2.1]heptan-2-ol   endo-bicyclo[2.2.1]heptan-2-ol

(Ketone reduction)

18. The acetyl chloride can be reduced first to ethanol using $LiAlH_4$. Acetyl chloride can also be reduced to acetaldehyde using H_2 on the surface of a $Pd/BaSO_4$ catalyst. These two products can then react with an acid catalyst to produce acetal. **(Acetal formation)**

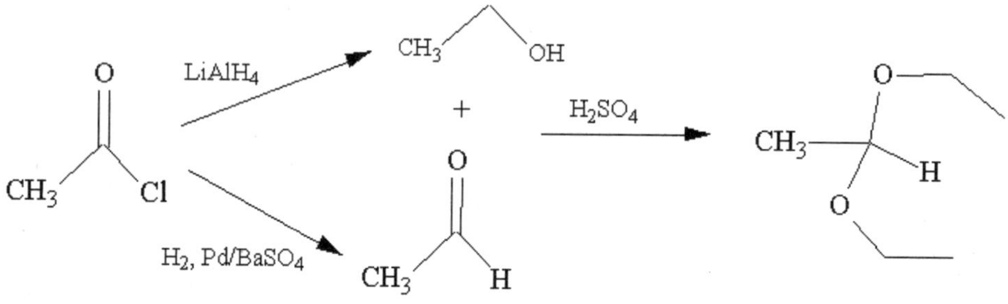

19. One way to do this synthesis is follows.

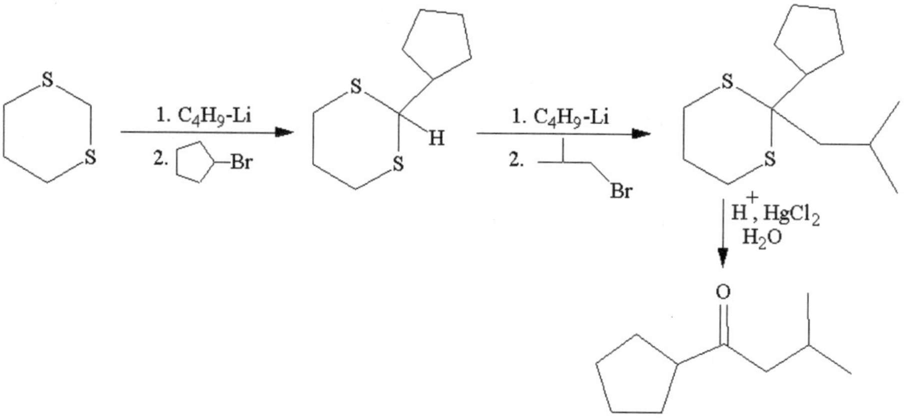

Another way is to reverse the order of adding bromocyclopentane and 1-bromo-2-methylpropane. **(Ketone synthesis)**

20. a. Acetyl chloride is converted to acetaldehyde by heating it with H_2 on the surface of a $Pd/BaSO_4$ catalyst. This is called the Rosenmund reduction. **(Aldehyde synthesis)**

21. a. Formaldehyde readily reacts with water to produce $CH_2(OH)_2$. The others are less reactive to water because alkyl groups decrease the reactivity of the carbonyl C atom. **(Hydration reactions)**

22. b. This compound is a cyclic hemiacetal. All hemiacetals have the general formula of R–CH(OH)(OR´). **(Hemiacetals)**

23. c. $(CH_3)_2NH$ will not react because it is a secondary amine. Primary amines and ammonia react with carbonyl compounds and produce imines. **(Imine formation)**

24. a. The product of the reaction is $C_6H_5–CH=NNH–C_6H_5$, benzaldehyde phenylhydrazone. **(Phenylhydrazone formation)**

25. d. The intermediate in enamine synthesis is a carbinolamine, $R–C(OH)(NR_2)–R'$. **(Enamine formation)**

26. e. The Tollens test is used to identify aldehydes by oxidizing them to acids and in the process producing a silver mirror in the reaction container. **(Aldehyde oxidation)**

27. d. CCl_3CHO reacts fastest of the listed compounds in cyanohydrin formation because the CCl_3 group has the greatest capacity to withdraw electron density from the carbonyl C atom of the listed compounds. The more positive the carbonyl C atom, the more reactive it is. **(Rates of nucleophilic substitution reactions)**

28. b. Isopropyl bromide can be converted to the necessary ylide which can react with benzaldehyde to produce the 2-methyl-1-phenyl-1-propene. **(Wittig reaction)**

29. a. In an acidic solution, $HOCH_2CH_2OH$ and $H_2C=O$ react and produce 1,3-dioxolane, a cyclic acetal. **(Acetal formation)**

30. b. $CH_3CH_2CH_2CH_2COCH_3$ is the compound because a negative Tollens test eliminates aldehydes, a positive iodoform test indicates that it is a methyl ketone, a positive semicarbazone shows that it is a ketone, and the formation of hexane upon reduction shows that the C chain is unbranched. **(Ketone identification)**

31. b. In the ^1H NMR spectrum, the chemical shift of the aldehydic proton, $–CHO$, is in the range of 9 to 10 ppm (δ). **(Spectroscopy of carbonyl compounds)**

32. c. $II < I < III < IV$ is the correct order. The greater the partial charge on the carbonyl C atom, the more the equilibrium shifts to the hydrate. Compound IV has a Cl atom adjacent to the carbonyl C atom; thus, it would have the greatest amount of hydrate in equilibrium. Aldehydes have less stabilization of the carbonyl C atom because there is only one attached R group. **(Hydrates)**

33. c. $CH_3CH(OD)CH_2CH_2CH_3$ is the product because the $NaBH_4$ produces the C–H bond and the D_2O forms the bond to the O atom. **(Reduction of carbonyl compounds)**

34. d. All of these pathways produce 2-hexanone. **(Ketone synthesis)**

35. d. First reacting with $(CH_3CH_2)_2Cd$ will produce 3-decanone, which is then reduced in the Wolff-Kishner reduction, H_2NNH_2, base, and heat, to decane. **(Synthesis of carbonyl compounds)**

Grade Yourself

Circle the number of questions you missed, then fill in the total incorrect for each topic. If you answered more than three questions incorrectly, you need to focus on that topic. (If a topic has less than three questions and you had at least one wrong, we suggest you study that topic also. Read your textbook, a review book, or ask your teacher for help.)

Subject: Aldehydes and Ketones

Topic	Question Numbers	Number Incorrect
Ketone nomenclature	1, 3	
Aldehyde nomenclature	2, 4	
Carbonyl structure	5	
Carbonyl compound synthesis	6, 35	
Aldehyde synthesis	7, 20	
Ketone synthesis	8, 19, 34	
Aldehyde reactions	9	
Hemiacetal formation	10	
Acetal formation	11, 18, 29	
Cyanohydrin formation	12	
Imine formation	13, 23	
Enamine formation	14, 25	
Semicarbazone formation	15	
Wittig reaction	16, 28	
Ketone reduction	17	
Hydration reactions	21	
Hemiacetals	22	
Phenylhydrazone formation	24	
Aldehyde oxidation	26	
Rates of nucleophilic substitution reactions	27	
Ketone identification	30	
Spectroscopy of carbonyl compounds	31	
Hydrates	32	
Reduction of carbonyl compounds	33	

Enols, Enolates, and Enamines

16

Brief Yourself

Aldehydes and ketones with an α H atom are in equilibrium with their enol forms, $RCH=C(OH)R'$.

$$RCH_2COR' \Leftrightarrow RCH=C(OH)R'$$

Enols result when a proton effectively transfers from an α C atom to the carbonyl O atom. This process is called keto-enol tautomerism. Actually, the proton comes from the solvent and not from an intramolecular transfer. For most common aldehydes and ketones the amount of enol present at equilibrium is small. The equilibrium constant, K, for the enolization of acetone is only on the order of magnitude of 10^{-9}. Either acids or bases can be added to carbonyl compounds to produce enols.

When base is added to a carbonyl compound with α H atoms, a resonance-stabilized enolate ion results.

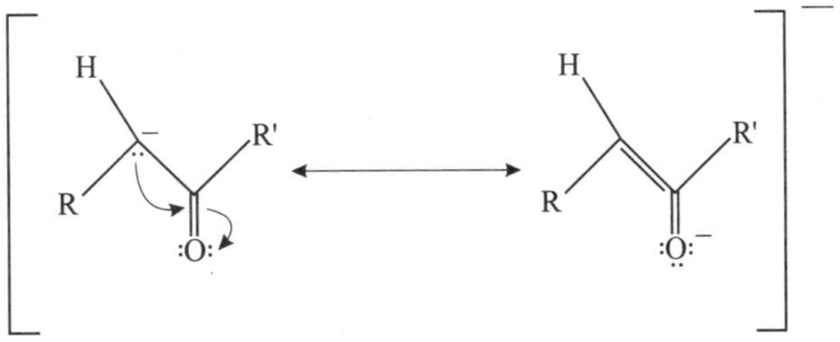

In this ion, the negative charge is dispersed from the α C atom to the carbonyl O atom. This delocalization is responsible for the acidity of the α H atoms in carbonyl compounds. The H atoms that are α to two carbonyl groups are even more acidic than those that are adjacent to just one carbonyl.

When the enolate ion forms in the presence of a halogen such as iodine, the α H atoms are replaced with I atoms. Methyl ketones, $RCOCH_3$, undergo a rapid reaction with halogens, producing a carboxylate anion and haloform, HCX_3.

$$RCOCH_3 + X_2(OH^-) \rightarrow RCOO^- + HCX_3$$

This is called the haloform reaction.

Aldehydes and ketones undergo addition-condensation reactions with themselves. This is called the aldol condensation reaction. In this reaction, the base removes a proton from the carbonyl compound and the resulting enolate ion attacks the carbonyl C atom as in a nucleophilic addition reaction.

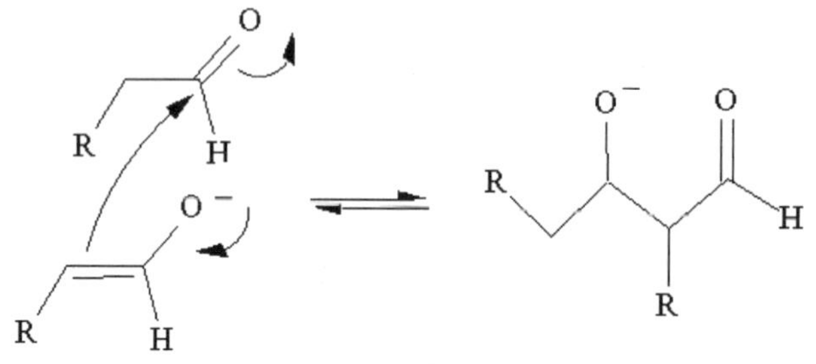

The resulting anion picks up a proton from the solvent, producing a β-hydroxy carbonyl compound. This compound when heated will eliminate water, producing α,β-unsaturated carbonyl compounds.

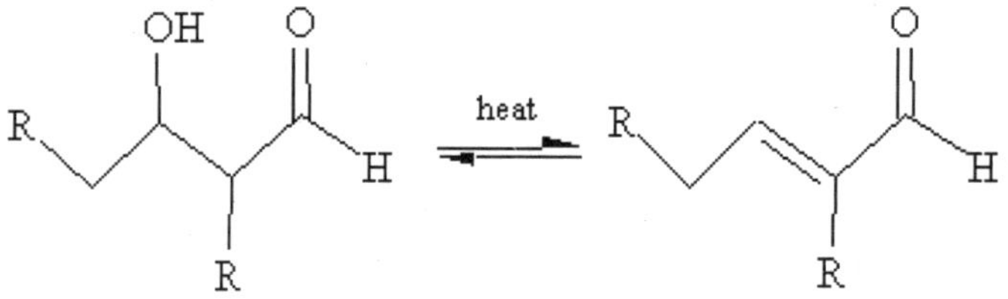

Most often aldol condensation reactions use one aldehyde or ketone, but if planned properly, a mixed aldol condensation can be carried out. In this reaction, two different carbonyl compounds are used. To obtain a single product, one of these compounds should not have H atoms on the α C atom.

The products of aldol condensation reactions, α,β-unsaturated carbonyl compounds, have a continuous system of π electrons. This means that α,β-unsaturated carbonyl compounds have electron delocalization over the C atoms in the C–C double bond and the carbonyl group. Because of this delocalization, these compounds have two electrophilic sites that can be attacked by nucleophiles.

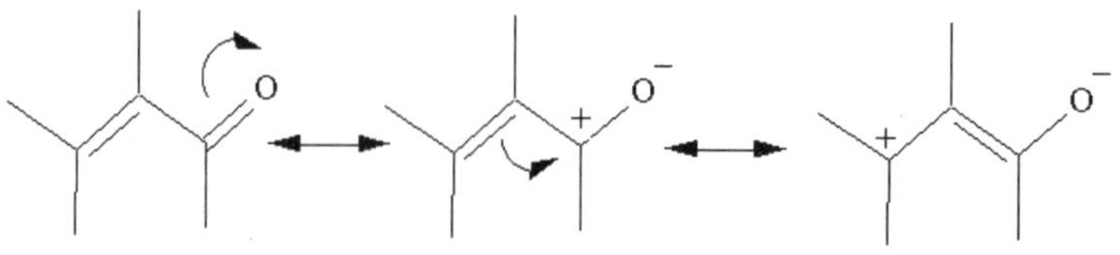

Thus, reagents either add directly across the carbonyl group, 1,2-addition, or add at the terminus of the double bond, 1,4-addition or conjugate addition. Strongly basic nucleophiles such as those found in Grignard reagents, organolithium compounds, and lithium aluminum hydride, tend to undergo 1,2-addition. Weakly basic nucleophiles tend to undergo conjugate addition.

Lithium dialkyl cuprates, LiCuR₂, in ether undergo conjugate addition to ,-unsaturated carbonyl compounds and produce β-alkyl substituted derivatives of the starting compound.

$$(CH_3)_2C=CHCOCH_3 + LiCu(CH_3)_2 \rightarrow (CH_3)_3CCH_2COCH_3$$

This is called an alkylation reaction. Another example of an alkylation reaction uses enolate ions. This is most commonly carried out with β-diketones because otherwise the same conditions needed for alkylation reactions will cause aldol condensation reactions. An example of such a reaction is as follows.

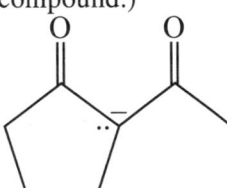

Another reaction that produces α-substituted carbonyl compounds is the reaction of pyrrolidine enamines with primary alkyl halides. These reactions produce α-substituted ketones and pyrrolidine after hydrolysis.

Test Yourself

1. a. What is an enol?

   b. Write an equation that shows the equilibrium of the enol and keto forms of 2-butanone.

   (For problems 2 to 4, consider the acid-catalyzed reaction of cyclopentanecarbaldehyde with Br₂ in chloroform.)

2. What product results from this reaction?

3. Write the complete mechanism for this reaction.

4. Show the resonance stabilization of the cation intermediate.

5. Draw structures of the keto and enol forms of 2-pentanone.

6. Draw the structures of the two most stable forms of 2,4-hexanedione.

(For problems 7 and 8, consider the following compound.)

7. Draw the structure of the enolate ion of this compound.

8. Show the resonance stabilization of the enolate ion from problem 7.

9. What are the products of the reaction of 3-methyl-2-hexanone with Br₂ in NaOH followed by the addition of acid?
   (For problems 10 and 11, consider the aldol condensation of propanal, CH₃CH₂CHO.

10. Draw the structure of the aldol addition product of propanal.

11. What product forms when the aldol addition product of propanal is heated in acid?

12. Draw the structure of the product of the aldol condensation reaction of acetophenone after dehydration.

13. What reactant(s) can undergo an aldol condensation to produce the following compound?

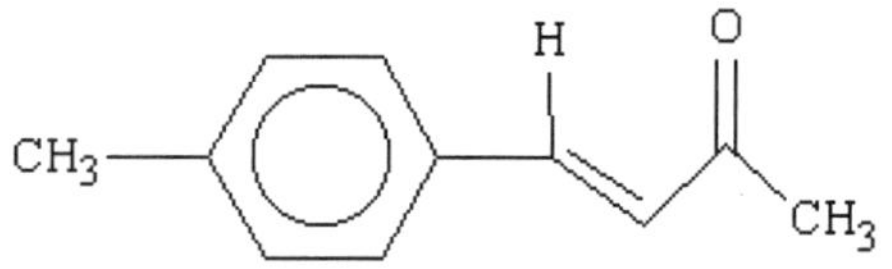

14. What product results when 2-phenylethanal undergoes an aldol condensation in base followed by dehydration?

15. Consider the following compound:

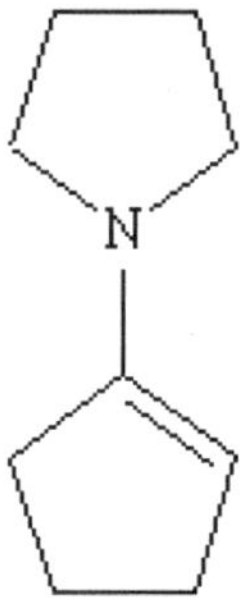

Draw the structure of the product that results when this compound reacts with benzyl bromide followed by acid hydrolysis.

16. Write an equation that shows the formation of the product when 2-methyl-1,3-cyclohexanedione reacts with benzyl bromide in a basic solution.

17. Starting with 2-methylpropanal, show how the following compound can be synthesized.

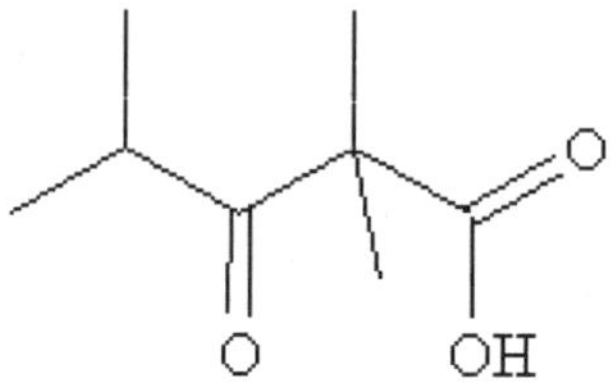

18. Draw the structure of the aldol condensation product of the following.

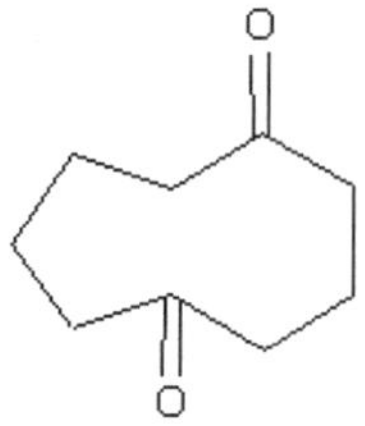

19. Consider the compound 2,7-octanedione.

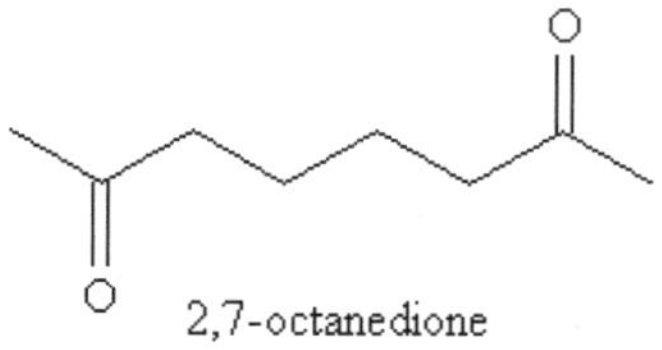

2,7-octanedione

2,7-Octanedione undergoes an aldol cyclization reaction. Write an equation that shows the addition and dehydration products of this reaction.

20. Consider the following molecule:

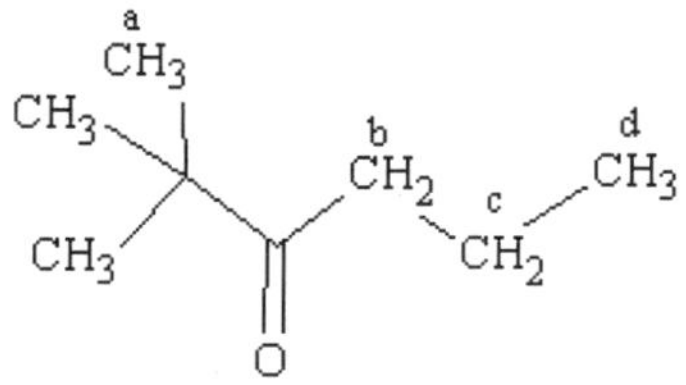

Which H atoms in this molecule are most acidic?

21. What is the product of the reaction of 2-methylbutanal with Br_2 and acetic acid?

    a. $CH_3CH_2CH(CH_3)COBr$

    b. $CH_3CH_2CBr(CH_3)CHO$

    c. $CH_3CHBrCH(CH_3)CHO$

    d. $CH_3CH_2CBr(CH_3)CBr(OH)$

    e. no reaction

22. What is the relationship of the following structures?

    CH₃COCH₃ and CH₃C(OH)=CH₂

    a. tautomers

    b. structural isomers

    c. enantiomers

    d. resonance structures

    e. no relationship

23. How many different enol forms of 2-methylcyclohexanone exist?

    a. zero

    b. one

    c. two

    d. three

24. Which of the following could **not** undergo self aldol condensation?

    a. diphenyl ketone

    b. 2,2-dimethylpropanal

    c. 2,2,4,4-tetramethyl-3-pentanone

    d. none of the above undergo self aldol condensation

25. Which of the following gives a positive iodoform test?

    a. cyclooctanone

    b. isobutyraldehyde

    c. acetophenone

    d. all of the above

26. What is the product of the reaction of 3-methyl-2-cyclohepten-1-one with LiCu(CH₃)₂ in ether followed by hydrolysis?

    a. 2,2-dimethylcycloheptanone

    b. 2,3-dimethylcycloheptanone

    c. 3,3-dimethylcycloheptanone

    d. 1,3-dimethylcycloheptanol

    e. none of these

27. Compound *A* reacts with methyl iodide in the presence of potassium carbonate and produces Compound *B*. *B* is reduced with LiAlH₄ to 3-methyl-2,4-pentanediol. What is the identity of *A*?

    a. CH₃COCH₂COCH₃

    b. CH₃COCH(CH₃)COCH₃

    c. CH₃CHBrCH₂CHBrCH₃

    d. CH₃COCH₂CH₂CH₃

    e. none of these

28. Compound *C* reacts with pyrrolidine to produce Compound *D*, which reacts with CH₃COCl to produce Compound *E*. *E* is hydrolyzed to the following compound.

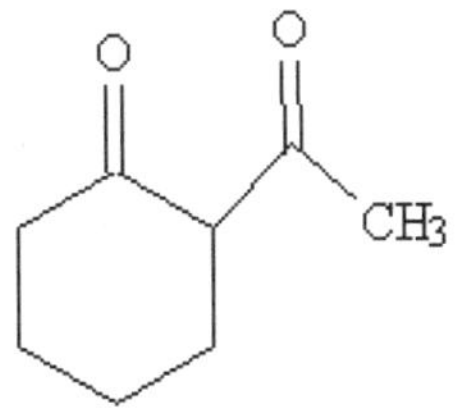

    What is the identity of *C?*

    a. CH₃COCH₃

    b. cyclohexene

    c. cyclohexanol

    d. cyclohexyl carbaldehyde

    e. none of these

29. Which of the following will undergo a base-catalyzed condensation reaction with cyclopentadiene to produce the following compound?

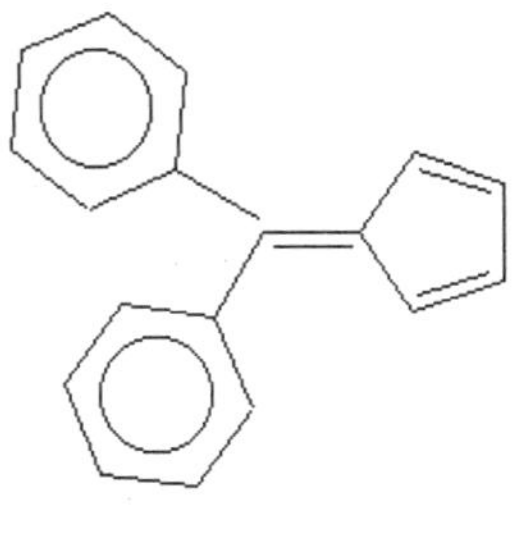

    a. benzaldehyde

    b. benzyl alcohol

    c. benzophenone

    d. acetophenone

    e. none of these

30. Starting with cyclohexanone, which of the
    following synthetic pathways will produce
    cyclohexane-1,1,3-d_3?

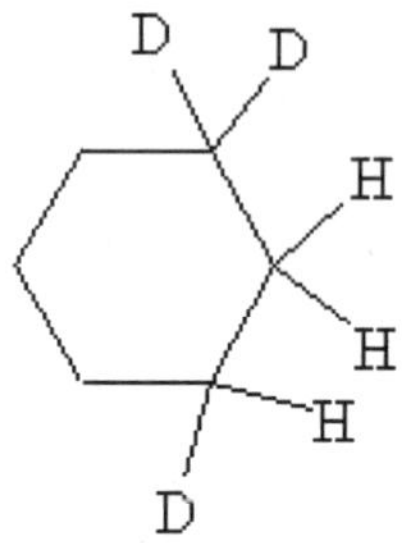

    a. 1. NaOD/D₂O, 2. NaBD₄

    b. 1. NaOD/D₂O, 2. H₂SO₄, heat, 3. H₂/Pt

    c. 1. NaOD/D₂O, 2. H₂SO₄, heat, 3. D₂/Pt

    d. 1. NaOD/D₂O, 2. Zn(Hg), HCl

    e. none of these

✔ Check Yourself

1. a. An enol has the general formula of $R_2C=CR'(OH)$.
    b. The equilibrium of 2-butanone with its enol form is as follows:

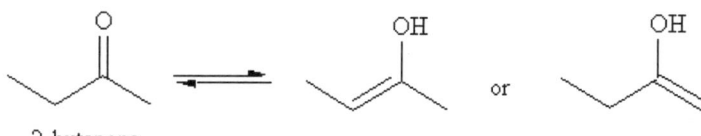

(Enols)

2. The product of this reaction is 1-bromocyclopentanecarbaldehyde because, under these conditions, a Br
    atom substitutes for the α H atom to the carbonyl.

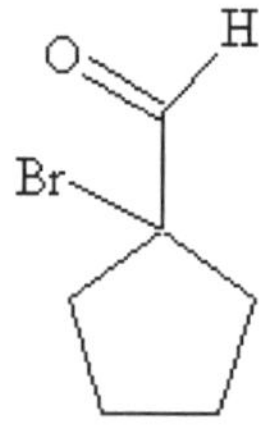

1-bromocyclopentanecarbaldehyde

(Halogenation of carbonyl compounds)

3. The first step of the mechanism, conversion to the enol form, is the rate-determining step.

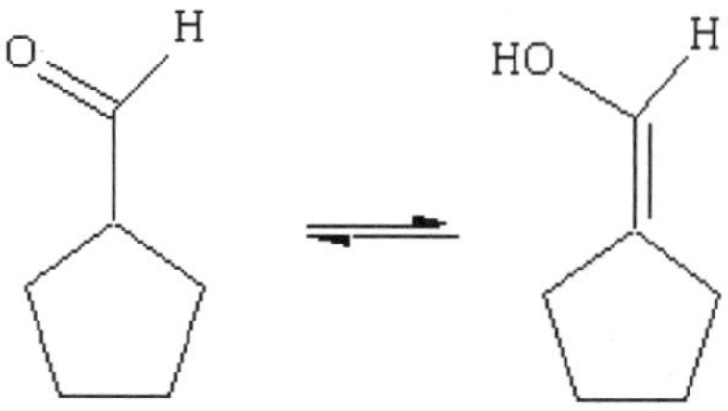

The second step of the mechanism is the reaction of the enol with the Br_2 to produce a resonance-stabilized cation intermediate.

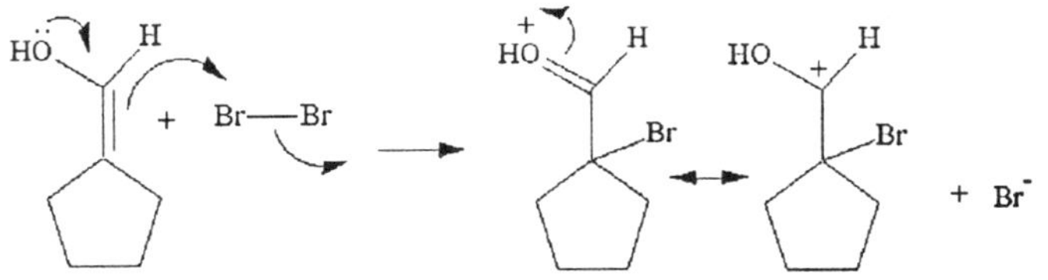

The final step of the mechanism occurs when the bromide ion accepts the proton from the protonated α-bromoketone.

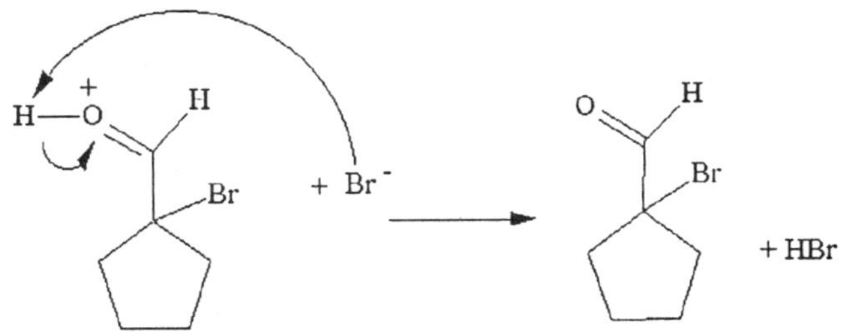

(Mechanism of halogenation)

4. The resonance stabilization of the intermediate cation can be shown as follows:

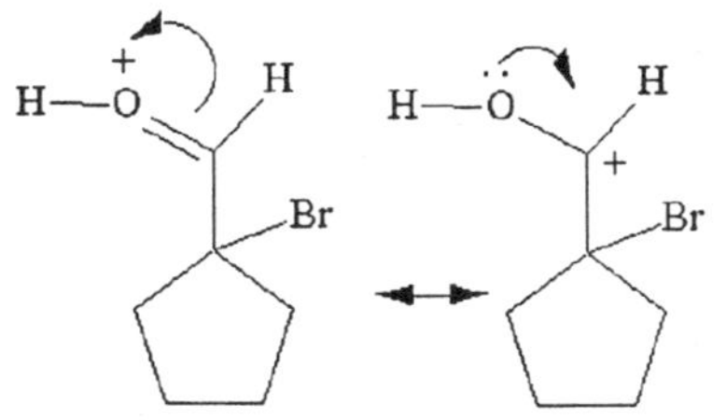

(Mechanism of halogenation)

5. The keto and enol forms of 2-pentanone are as follows:

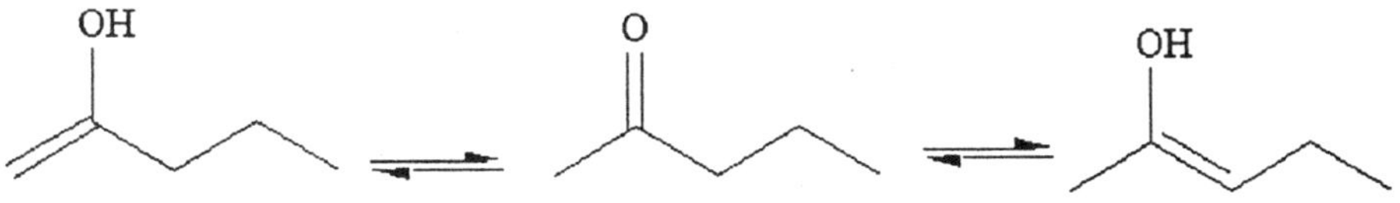

(Keto and enol forms)

6. The two most stable forms of 2,4-hexanedione result when the double bond forms between the two carbonyl groups. These enols are stabilized by resonance and intramolecular hydrogen bonding.

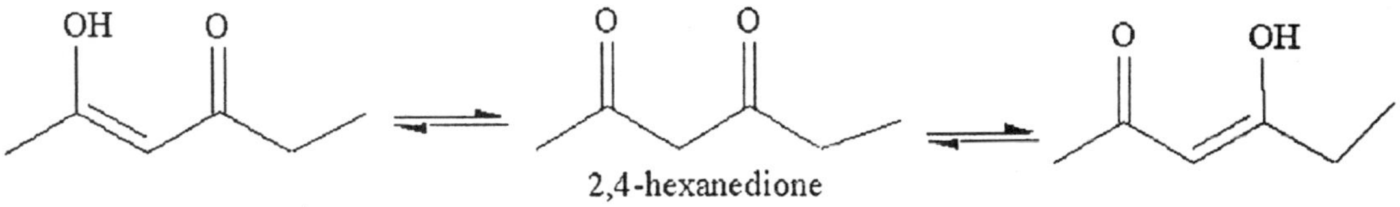

2,4-hexanedione

(Stability of enols)

7. The following shows the enolate ion that results:

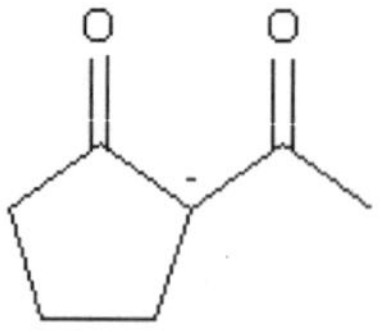

(Enolate ions)

8. There are three principal resonance structures for this enolate ion.

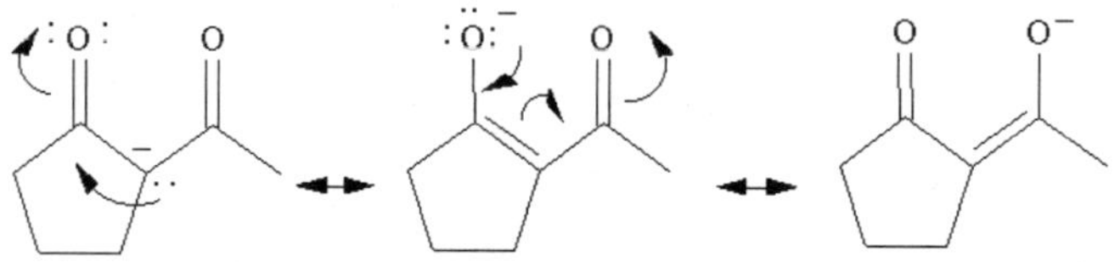

(Enolate ions)

9. The reaction that takes place is the bromoform reaction in which an acid and bromoform, $CHBr_3$, result. The acid that results is 2-methylpentanoic acid, $CH_3CH_2CH_2CH(CH_3)COOH$. **(Bromoform reaction)**

10. The aldol addition product of propanal is 3-hydroxy-2-methylpentanal.

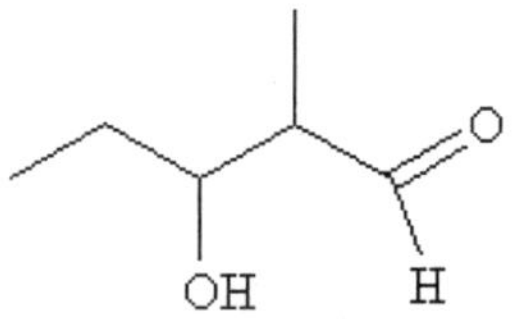

3-hydroxy-2-methylpentanal

(Aldol addition)

11. When aldol addition products are heated, they dehydrate and form α,β-unsaturated carbonyl compounds.

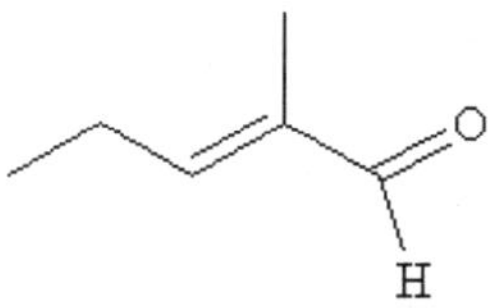

2-methyl-2-pentenal

Both *E* and *Z* isomers will form. **(Aldol condensation)**

12. The product of the aldol condensation of acetophenone is 1,3-diphenyl-2-buten-1-one.

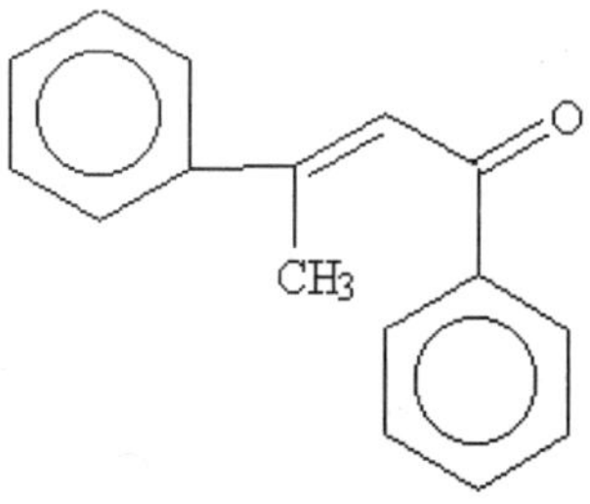

1,3-diphenyl-2-buten-1-one

(Aldol condensation)

13. The two reactants needed to produce the desired α, β-unsaturated compound are as follows.

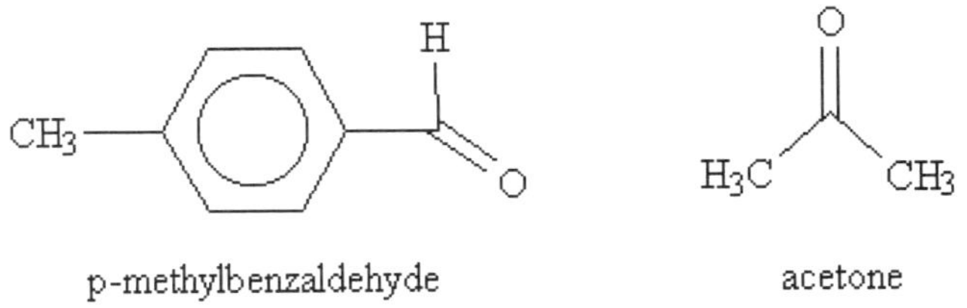

p-methylbenzaldehyde   acetone

(Mixed aldol condensations)

14. Two molecules of 2-phenylethanal, $C_6H_5-CH_2CHO$, react in an aldol condensation and produce 3-hydroxy-2,4-diphenylbutanal, $C_6H_5-CH_2CH(OH)CH(C_6H_5)CHO$. This product dehydrates to the following compound.

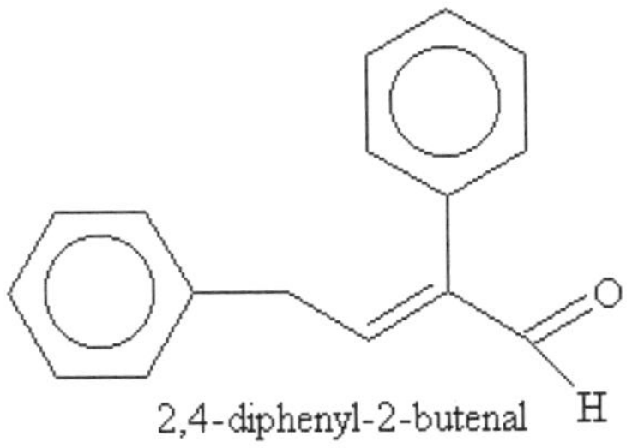

2,4-diphenyl-2-butenal   H

(Aldol condensation)

15. Enamines displace the Br atom in benzyl bromide producing an alkylated iminium salt which upon hydrolysis produces an alkylated ketone.

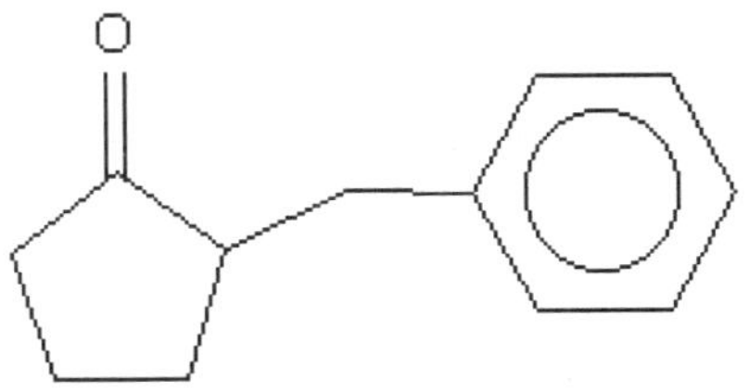

(Enamine synthesis)

16. In this reaction, alkylation occurs on the C atom to both α carbonyl groups.

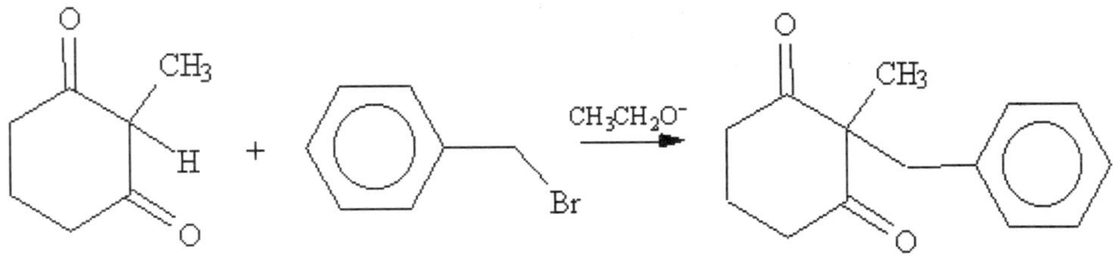

(α-Alkylation of carbonyl compounds)

17. The synthesis begins by adding base to produce the aldol addition product, which is then oxidized to the desired product.

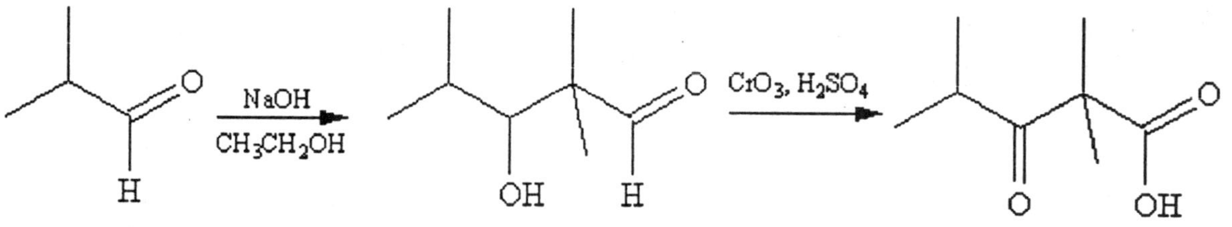

(**Aldol condensation**)

18. The aldol condensation product for this reaction is as follows:

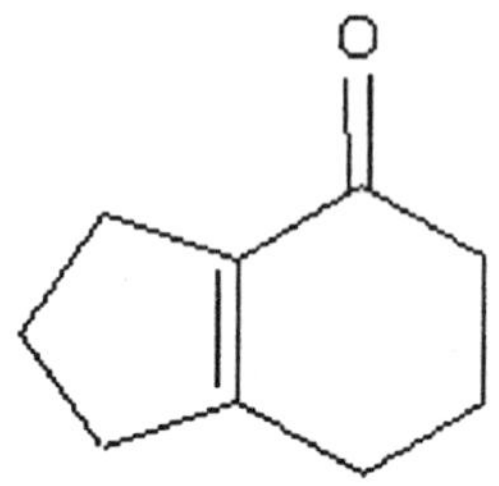

(**Aldol condensation**)

19. The reaction takes place as follows:

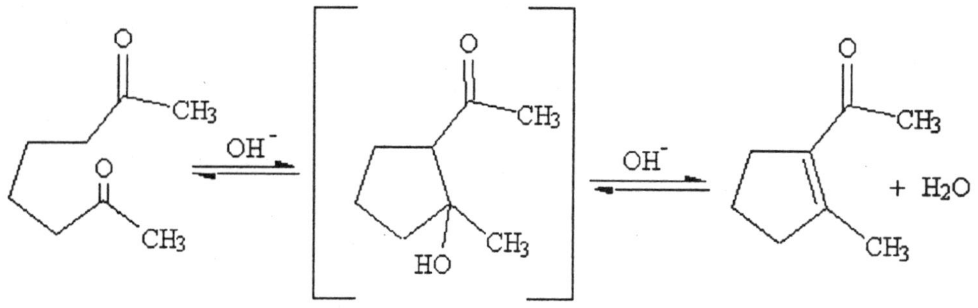

(**Aldol cyclization**)

20. b. The most acidic H atoms, the ones that will be most readily removed by a base, are those α to the carbonyl C atom. (α **-Hydrogen atoms**)

21. b. $CH_3CH_2CBr(CH_3)CHO$ is the product because under acidic conditions Br atoms substitute at the α position to the carbonyl group. (**Alpha substitution**)

22. a. These molecules are tautomers, isomers that can quickly interconvert by the movement of a proton from one site to another. (**Tautomerism**)

23. c. The two enol forms of 2-methylcyclohexanone are 2-methyl-1-cyclohexenol and 6-methyl-1-cyclohexenol. (**Keto-enol tautomerism**)

24. d. None of these compounds undergo self aldol condensation because they do not have α -hydrogen atoms that can be removed. (**Aldol condensation**)

25. c. Acetophenone is the only one that gives a positive iodoform test because it is the only methyl ketone. (**Haloform reaction**)

26. c. 3,3-Dimethylcycloheptanone results because lithium dimethylcuprate undergoes conjugate addition to α,β-unsaturated carbonyl compounds. **(Conjugate addition)**

27. a. Compound *A* is $CH_3COCH_2COCH_3$. It reacts with CH_3I and produces $CH_3COCH(CH_3)COCH_3$, which is reduced with $LiAlH_4$ to 3-methyl-2,4-pentanediol. **(Alkylation reactions)**

28. e. Compound *C* is cyclohexanone, which reacts with pyrrolidine to form an enamine of cyclohexanone. Acetyl chloride reacts at the α position, releasing HCl and producing the enamine derivative of the desired product. **(Enamine synthesis)**

29. c. Benzophenone reacts with cyclopentadiene under basic conditions and then eliminates water to produce the product. **(Condensation reaction)**

30. Treatment with NaOD and D_2O will produce cyclohexanone-2,2,5,5-*d4*, which undergoes dehydration when treated with sulfuric acid and heated. The desired product, cyclohexane-1,1,3-*d3*, is then produced when it is catalytically hydrogenated. **(Enolization)**

Grade Yourself

Circle the number of questions you missed, then fill in the total incorrect for each topic. If you answered more than three questions incorrectly, you need to focus on that topic. (If a topic has less than three questions and you had at least one wrong, we suggest you study that topic also. Read your textbook, a review book, or ask your teacher for help.)

Subject: Enols, Enolates, and Enamines

Topic	Question Numbers	Number Incorrect
Enols	1	
Halogenation of carbonyl compounds	2	
Mechanism of halogenation	3, 4	
Keto and enol forms	5	
Stability of enols	6	
Enolate ions	7, 8	✔
Bromoform reaction	9	✔
Aldol addition	10	
Aldol condensation	11, 12, 14, 17, 18, 24	
Mixed aldol condensations	13	
Enamine synthesis	15, 28	✔
α-Alkylation of carbonyl compounds	16	
Aldolcyclization	19	
α-Hydrogen atoms	20	
Alpha substitution	21	
Tautomerism	22	
Keto-enol tautomerism	23	
Haloform reaction	25	
Conjugate addition	26	
Alkylation reactions	27	
Condensation reaction	29	✔
Enolization	30	

Carboxylic Acids

 **Brief Yourself**

Carboxylic acids, also called organic acids, have the general formula of RCOOH. Each carboxylic acid contains a carboxyl group, –COOH. The geometry of the carboxyl group is trigonal planar because the central C atom is sp^2 hybridized.

The bond angles at the central C atom in the carboxyl group are near $120°$.

To name carboxylic acids, the *e* is removed from the parent alkane and replaced with *oic acid*. The simplest carboxylic acid is methanoic acid, HCOOH, most commonly called formic acid. When the carboxyl group is bonded to a ring, then *carboxylic acid* is added to the name of the ring. For example, the following is cyclopentanecarboxylic acid.

cyclopentanecarboxylic acid

If two carboxyl groups are found in a molecule, then *dioic acid* is added to the name of the parent alkane. For example, $HOOCCH_2COOH$ is the molecule propanedioic acid, commonly called malonic acid.

Due to the C=O and OH group in acids, they can form strong hydrogen bonds with themselves or with solvents in solution. Because of the strong hydrogen bonding, carboxylic acids have higher melting and boiling points than alcohols, aldehydes, and ketones with similar molecular masses. All of the carboxylic acids with fewer than five C atoms are miscible with water.

Carboxylic acids are the most acidic of the common organic compounds. Their acidity results from their ability to produce a resonance-stabilized carboxylate anion.

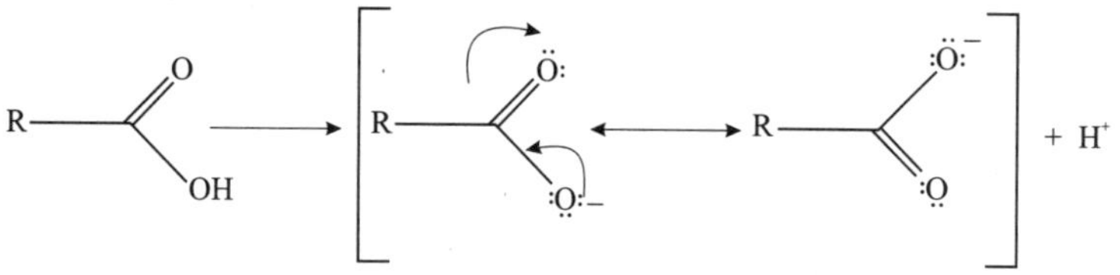

Additionally, the polar carboxylate anion is stabilized by hydration with water molecules. The addition of substituent groups to acids increases their acidity if they help to disperse the negative charge in the carboxylate anion and decrease the acidity if they intensify the negative charge. Dicarboxylic acids tend to be stronger than monocarboxylic acids because the second carboxyl group helps to stabilize the carboxylate anion and there is a greater statistical chance of losing a proton. Carboxylic acids, as with all acids, are neutralized by bases and produce salts.

Carboxylic acids can be prepared by oxidative cleavage of alkenes using hot concentrated $KMnO_4$. Actually the potassium salt of the acid results, which then must be hydrolyzed to the acid.

$$RCH=CHR' + KMnO_4 \text{ (heat)} \rightarrow RCOOH + R'COOH$$

Primary alcohols and aldehydes can be oxidized by strong oxidizing agents to acids.

$$RCH_2OH + K_2Cr_2O_7, \ H^+ \rightarrow RCOOH$$

$$RCHO + K_2Cr_2O_7, \ H^+ \rightarrow RCOOH$$

Alkyl-substituted benzene compounds can be oxidized to benzoic acid, C_6H_5–COOH. Methyl ketones react with halogens and base, followed by hydrolysis, to produce carboxylic acids with one less C atom than the original ketone. Additionally, haloform, CHX_3, is released.

$$RCOCH_3 + I_2(OH^-) \rightarrow RCOOH + CHI_3$$

Carboxylic acids are also synthesized by reacting a Grignard reagent with carbon dioxide.

$$RMgX + CO_2 \rightarrow RCO_2MgX + H^+(aq) \rightarrow RCOOH$$

Note that in the Grignard synthesis of an acid, the resulting acid has one more C atom than the initial Grignard reagent. This is also true of preparing carboxylic acids through the acid hydrolysis of nitriles, RCN.

$$RBr + CN^- \rightarrow RCN + H^+(aq) \rightarrow RCOOH$$

Carboxylic acids can be reduced with strong reducing agents such as LiAlH$_4$ to primary alcohols.

$$RCOOH + LiAlH_4 \rightarrow RCH_2OH$$

When carboxylic acids react with thionyl chloride, SOCl$_2$, they produce acid chlorides.

$$RCOOH + SOCl_2 \rightarrow RCOCl$$

Acid chlorides react with alcohols to produce esters.

$$RCOCl + R'OH \rightarrow RCOOR' + H_2O$$

Acid chlorides can be reduced to aldehydes by using weak reducing agents such as lithium aluminum tri(t-butoxy)hydride, LiAl[OC(CH$_3$)$_3$]$_3$H.

$$RCOCl + LiAl[OC(CH_3)_3]_3H \rightarrow RCHO$$

Hydroxy carboxylic acids in which the OH group is on the fourth or fifth C atom will undergo an intramolecular esterification reaction and produce lactones, cyclic esters with five- and six-membered rings, respectively.

Carboxylic acids can be α-halogenated by treating them with a halogen, X$_2$, and PCl$_3$ or P as a catalyst.

$$RCH_2COOH + X_2 \rightarrow RCHXCOOH + HX$$

This is called the Hell-Volhard-Zelinsky reaction.

The carboxyl group of acids can be removed in the Hunsdiecker reaction. In this reaction, the silver salt of the acid is treated with a halogen, X$_2$, and heated. The products of this reaction are an alkyl halide, silver halide, and carbon dioxide.

$$RCOO^- Ag^+ + X_2 \rightarrow RX + AgX + CO_2$$

When 1,3-dicarboxylic acids are heated they also decarboxylate, releasing CO$_2$.

$$RCH(COOH)_2 + heat \rightarrow RCH_2COOH + CO_2$$

Test Yourself

1. a. What is the functional group in carboxylic acids?

   b. Write a paragraph that describes the structure of this functional group.

2. What are the IUPAC and common names of the following carboxylic acid?

3. Draw the structure of (*E*)-2-methyl-3-phenyl-2-propenoic acid.

4. Succinic acid is a dicarboxylic acid with an unbranched chain of four C atoms.

    a. Draw the structure of succinic acid.

    b. What is the IUPAC name of succinic acid?

5. Draw a structure that shows the intermolecular hydrogen bonding found in acetic acid.

6. Which of the following is a stronger acid? Explain your answer.

7. List two ways that carboxylic acids can be prepared through an oxidation reaction. Write general equations for each of these oxidations.

8. How could 4-hexen-2-one be converted to 3-pentenoic acid? Write an equation for this reaction.

9. Starting with 3-bromo-1-propene, outline a synthesis of glutaric acid, $HOOC(CH_2)_3COOH$.

10. Starting with *t*-butyl bromide, outline a synthesis of 2,2-dimethylpropanoic acid.

11. In the first step of the mechanism for the acid-catalyzed esterification reaction, the carboxylic acid is protonated by the acid catalyst. Write an equation for this step, showing specifically which of the two O atoms is protonated. Fully explain your answer.

12. a. What product results when 4-hydroxy-2-methylpentanoic acid undergoes an intramolecular esterification reaction?

    b. In what general class of compounds does the product belong?

13. Draw the structure of the final product that results when acetic acid reacts with Br_2 in the presence of PCl_3 and this compound is treated with CN^- followed by hydrolysis.

14. Malonic acid, $HOOCCH_2COOH$, undergoes a decarboxylation reaction when heated. Write a mechanism that accounts for this reaction.

15. *o*-Methyltoluene (*o*-xylene) is oxidized with chromic acid to Compound *A*. *A* is treated with ethanol and acid, producing Compound *B*. Draw the structure of *B*.

16. Compounds *C* and *D* are cyclic alkene dicarboxylic acids that have carboxyl groups on adjacent C atoms. When *C* and *D* are heated with potassium permanganate in base followed by acid hydrolysis, the following compound results in both reactions. Draw the structures of *C* and *D*.

17. Compound *E* undergoes acid hydrolysis releasing NH_3 and produces Compound *F*. *F* reacts with Br_2 and PBr_3 and produces Compound *G*. When *G* is treated with KOH and alcohol it produces the following compound.

    Draw the structures of *E*, *F*, and *G*.

18. By what method could heptanoic acid be converted to heptanal?

19. Which of the isomeric fluorobenzoic acids is the strongest acid? Explain.

20. Which of the following has the highest boiling point?

    a. $CH_3CH_2CH_2CH_3$

    b. $CH_3CH_2CH_2OH$

    c. CH_3CH_2CHO

    d. CH_3COOH

21. Which of the following is the strongest acid?

    a. CH_3COOH

    b. $ClCH_2COOH$

    c. $Cl_2CHCOOH$

    d. Cl_3CCOOH

22. Which of the following is the strongest acid?

    a. CH_3COOH

    b. $HOOCCH_2COOH$

    c. $HOOCCH_2CH_2COOH$

    d. $HOOC(CH_2)_4COOH$

23. Which of the following will most readily undergo acid-catalyzed esterification with benzoic acid?

    a. $CH_3CH_2CH_2OH$

    b. $CH_3CH_2CH(OH)CH_3$

    c. $(CH_3)_2C(OH)CH_2CH_3$

    d. $(CH_3)_3COH$

24. What is the product of the reaction of $CH_3CH_2{}^{18}OH$ with CH_3COOH in the presence of HCl?

    a. $CH_3COOCH_2CH_3 + H_2{}^{18}O$

    b. $CH_3CO^{18}OCH_2CH_3 + H_2O$

    c. $CH_3COOCH_2CH_3 + H_2O$

    d. none of these

25. Which of the following compounds most readily forms a lactone?

    a. *p*-hydroxybenzoic acid

    b. 4-hydroxyhexanoic acid

    c. 2-hydroxyhexanoic acid

    d. 6-hydroxyhexanoic acid

    e. none of these

26. Which of the following compounds undergoes a decarboxylation reaction?

    a. 1,1-cyclopentanedicarboxylic acid

    b. 2-ethylmalonic acid

    c. 2,2-diphenylpropanedioic acid

    d. all of these

27. Compound *X* undergoes oxidation with hot concentrated $KMnO_4$, followed by hydrolysis to give the following compound.

    What is the identify of *X?*

    a. cyclohexene

    b. 1,2-dimethylcyclohexene

    c. 3,6-dimethylcyclohexene

    d. 2,4-dimethylcyclohexene

    e. none of these

28. Which of the following compounds could be used to begin a Grignard synthesis of 3-methylbutanoic acid?

    a. 1-bromo-2-methylpropane

    b. 2-bromo-2-methylpropane

    c. 1-bromo-2-methylbutane

    d. none of these

29. Which of the following is most commonly used to convert carboxylic acids to acid chlorides?

    a. Cl_2

    b. SCl_2, HCl

    c. $SOCl_2$

    d. SO_2Cl_2

    e. none of these

30. Which of the following is **incorrect** about the spectra of carboxylic acids?

    a. In the IR spectrum, the O–H stretch of the carboxyl group is a broad absorbance in the range of 2700 to 3500 cm^{-1}.

    b. In the ^{1}H NMR spectrum, the proton on the OH group typically is in the range of 10 to 12 ppm (δ).

    c. In the ^{13}C NMR spectrum, the carbonyl C atom is usually in the range of 140 to 150 ppm.

    d. In the mass spectrum, the molecular ion peak is usually small because there are many ways the acid can fragment

✔ Check Yourself

1. a. The functional group in carboxylic acids is the carboxyl group, –COOH.
    b. Because of the C–O double bond in the carboxyl group, the C and O atoms are sp^2 hybridized; hence, its geometry is trigonal planar with bond angles near 120°. **(Carboxylic acids)**

2. The common name for this acid is salicylic acid. Its IUPAC name is 2-hydroxybenzenecarboxylic acid. **(Acid nomenclature)**

3. The structure of (*E*)-2-methyl-3-phenyl-2-propenoic acid is as follows.

    (Acid structure)

4. a. The structure of succinic acid is HOOCCH$_2$CH$_2$COOH.
    b. Its IUPAC name is butanedioic acid. **(Acid nomenclature)**

5. Two molecules of acetic acid hydrogen bond with each other as follows.

    The partial positive charge on the H atoms attracts the partial negative charge on the O atoms. **(Properties of carboxylic acids)**

6. 4-Nitrobenzoic acid is a stronger acid ($K_a = 3.8 \times 10^{-4}$) than benzoic acid ($K_a = 6.3 \times 10^{-5}$) because the nitro group withdraws electron density from the carboxylate anion. Dispersal of charge in the carboxylate anion shifts the equilibrium to the right. **(Acidity of acids)**

7. Both primary alcohols and aldehydes can be oxidized to carboxylic acids.

    $RCH_2OH + K_2Cr_2O_7(H_2SO_4) \rightarrow RCOOH$

    $RCHO + K_2Cr_2O_7(H_2SO_4) \rightarrow RCOOH$

    (Synthesis of acids)

8. The iodoform reaction cleaves the terminal C atom from methyl ketones and produces iodoform, CHI_3, and a carboxylic acid. **(Synthesis of acids)**

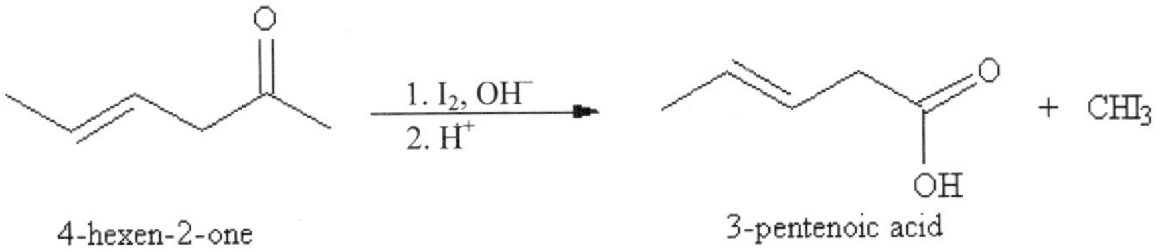

    4-hexen-2-one   3-pentenoic acid

9. Begin by reacting 3-bromo-1-propene with HBr in the presence of peroxides to give 1,3-dibromopropane.

    $BrCH_2CH=CH_2 + HBr \text{ (peroxides)} \rightarrow BrCH_2CH_2CH_2Br$

    Next, remove the two Br groups by doing an S_N2 displacement with aqueous CN^-.

    $BrCH_2CH_2CH_2Br + CN^-(aq) \rightarrow NCCH_2CH_2CH_2CN$

    Finally, hydrolyze the dinitrile with acid to glutaric acid.

    $NCCH_2CH_2CH_2CN + H^+ \rightarrow HOOCCH_2CH_2CH_2COOH$

    (Synthesis of acids)

10. Substitution of the Br with a CN^- followed by hydrolysis is not possible because this is a tertiary halide. The best method to accomplish the conversion is through a Grignard synthesis using CO_2.
    (Synthesis of acids)

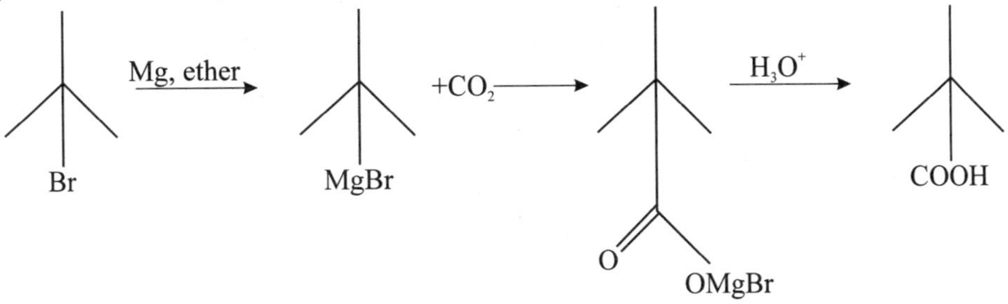

11. The first step of the acid-catalyzed esterification reaction is the protonation of the carbonyl O. The carbonyl O atom is protonated because the resulting cation is resonance stabilized.

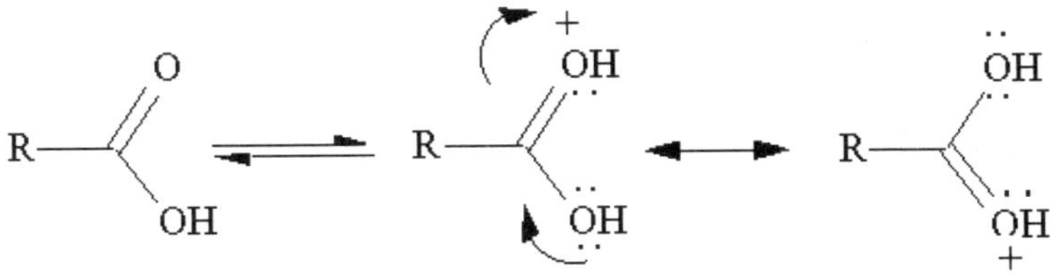

    There is no resonance stabilization if the hydroxyl O atom is protonated. **(Esterification reaction)**

12. a. The intramolecular esterification reaction of 4-hydroxy-2-methylpentanoic acid is as follows.

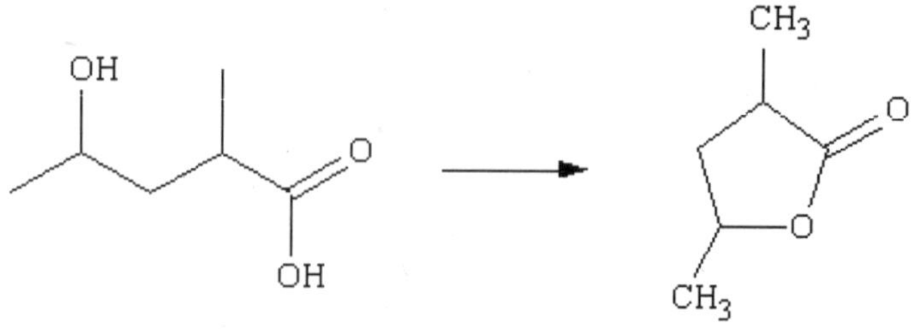

 b. A cyclic ester is a lactone. (**Esterification reaction**)

13. Acetic acid reacts with Br_2 and PCl_3 and produces bromoacetic acid, $BrCH_2COOH$. This is the Hell-Volhard-Zelinsky reaction. Treatment with CN^- followed by hydrolysis produces $HOOCCH_2COOH$, malonic acid. (**Hell-Volhard-Zelinsky reaction**)

14. A decarboxylation reaction is one where CO_2 is eliminated; thus, malonic acid is converted to acetic acid and CO_2 in this reaction. The following is a mechanism that explains this change.

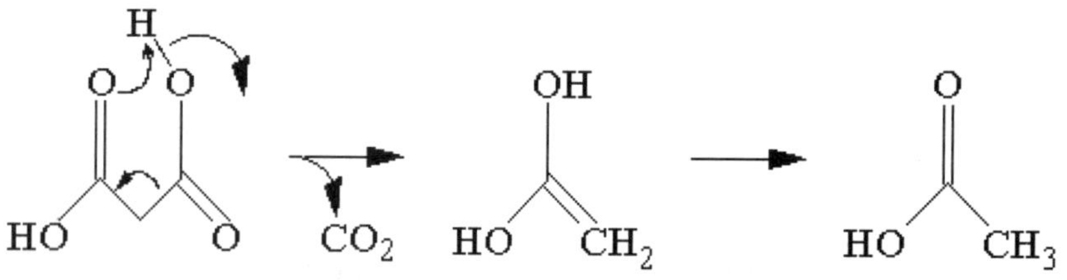

 Initially, the enol form of acetic acid results from this concerted mechanism. It then tautomerizes to the keto form of acetic acid. (**Decarboxylation reaction**)

15. *o*-Methyltoluene, also called *o*-xylene, is oxidized to 1,2-benzenedioic acid (*A*), also called phthalic acid, which reacts with ethanol to produce the ethyl ester of phthalic acid. (**Acid reactions**)

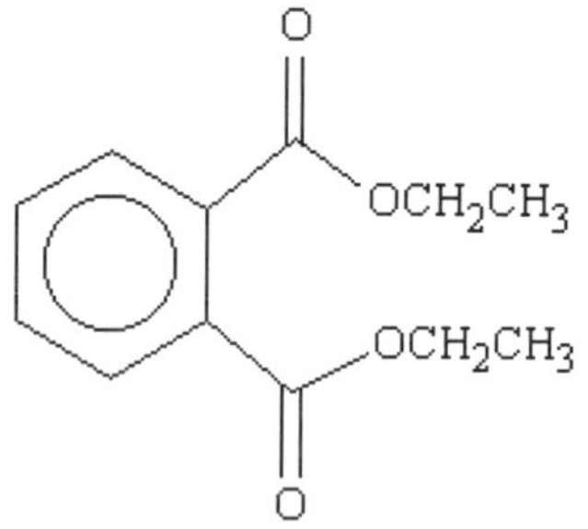

16. Alkenes are oxidatively cleaved to 1,2-dicarboxylic acids; thus, at the position between the two carboxyl groups, Compounds *C* and *D* have a C–C double bond. Hence, the structures of these molecule are as follows. (**Acid reactions**)

17. Compound *E* is cyanocyclohexane, C_6H_{11}–CN. Compound *F* is cyclohexane carboxylic acid, C_6H_{11}–COOH. Compound *G* is 1-bromocyclohexane carboxylic acid. (**Acid reactions**)

18. Heptanoic acid must first be converted to an acid chloride using $SOCl_2$.

    $CH_3CH_2CH_2CH_2CH_2CH_2COOH + SOCl_2 \rightarrow CH_3CH_2CH_2CH_2CH_2CH_2COOCl + SO_2 + HCl$

    The acid chloride is then reduced to the aldehyde using the bulky reducing agent lithium aluminum tri(*t*-butoxy)hydride, $LiAl[OC(CH_3)_3]_3H$.

    $CH_3CH_2CH_2CH_2CH_2CH_2COOCl + LiAl[OC(CH_3)_3]_3H \rightarrow CH_3CH_2CH_2CH_2CH_2CH_2CHO$

    (**Synthesis of acids**)

19. Of the three isomeric fluorobenzoic acids, *o*-fluorobenzoic acid is the strongest. F atoms can only stabilize the carboxylate anion through inductive effects; thus, the isomer where the F atom is closest to the carboxylate group is the one that should produce the greatest effect. The pK_a of *o*-fluorobenzoic acid is smaller, 3.3 (smaller pK_a means a stronger acid), than either *m*-fluorobenzoic acid, 3.9, or *p*-fluorobenzoic acid, 4.1. (**Acidity of acids**)

20. d. CH_3COOH has the highest boiling point because of its strong hydrogen bonding. (**Properties of carboxylic acids**)

21. d. Trichloroacetic acid, Cl_3CCOOH, is the strongest acid because the three Cl atoms help disperse the charge in the carboxylate anion. Three Cl atoms can do this more effectively that two, one, or no Cl atoms. (**Acidity of acids**)

22. b. Oxalic acid, $HOOCCH_2COOH$, is the strongest acid. Short-chain dicarboxylic acids are stronger acids because the second carboxyl group helps stabilize the carboxylate anion. Also, they have a greater probability of donating a proton than monocarboxylic acids. (**Acidity of acids**)

23. a. $CH_3CH_2CH_2OH$ is the most reactive because it is a primary alcohol which forms a less sterically hindered sp^3 intermediate. (**Esterification reaction**)

24. b. $CH_3CO^{18}OCH_2CH_3 + H_2O$ is the correct answer because in the acid-catalyzed esterification reaction the alcohol attacks the carbonyl C atom; thus, the ^{18}O-labeled atom will become part of the ester, specifically the one bonded to the ethyl group. (**Esterification reaction**)

25. b. 4-Hydroxyhexanoic acid most readily forms a lactone, a cyclic ester, because it produces a five-membered ring. The others either do not form an ester or the ring size would be too large or small. (**Lactone formation**)

26. d. All of these compounds undergo a decarboxylation reaction because they each have two carboxyl groups separated by a C atom. (**Decarboxylation reaction**)

27. c. *X* is 3,6-dimethylcyclohexene, which undergoes oxidative cleavage to the given dicarboxylic acid. (**Synthesis of acids**)

28. a. 1-Bromo-2-methylpropane, $CH_3CH(CH_3)CH_2Br$, first reacts with Mg in ether and yields a Grignard reagent. The Grignard reagent then reacts with CO_2 followed by hydrolysis to give 3-methylbutanoic acid. (**Synthesis of acids**)

29. c. Thionyl chloride, $SOCl_2$, is one of the most common reagents used to convert acids to acid chlorides. (**Acid reactions**)

30. c. In the ^{13}C NMR spectrum, the carbonyl C atom is usually in the range of 160 to 180 ppm. (**Spectra of acids**)

Grade Yourself

Circle the number of questions you missed, then fill in the total incorrect for each topic. If you answered more than three questions incorrectly, you need to focus on that topic. (If a topic has less than three questions and you had at least one wrong, we suggest you study that topic also. Read your textbook, a review book, or ask your teacher for help.)

Subject: Carboxylic Acids

Topic	Question Numbers	Number Incorrect
Carboxylic acids	1	
Acid nomenclature	2, 4	
Acid structure	3	
Properties of carboxylic acids	5, 20	
Acidity of acids	6, 19, 21, 22	
Synthesis of acids	7, 8, 9, 10, 18, 27, 28	
Esterification reaction	11, 12, 23	
Hell-Volhard-Zelinsky reaction	13	
Decarboxylation reaction	14, 26	
Acid reactions	15, 16, 17, 29	✓
Lactone formation	25	
Spectra of acids	30	

Carboxylic Acid Derivatives

18

Brief Yourself

The most important acid derivatives are acyl halides, RCOX; acid anhydrides, RCOOCOR′; esters, RCOOR′; and amides, $RCONH_2$, RCONHR′, and RCONR′$_2$. Additionally, nitriles, RCN, are also considered acid derivatives because they can be hydrolyzed to carboxylic acids.

The names of acyl halides are derived from the names of the corresponding acyl group followed by the halide name. For example, the following compound is 2-methyl-2-hexenoyl chloride:

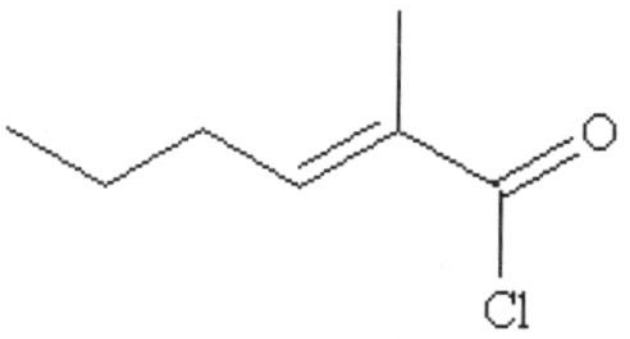

2-methyl-2-hexenoyl chloride

The names of carboxylic acid anhydrides are derived from the names of the acid that produced the anhydride. The word *anhydride* replaces *acid* in the names of anhydrides. For example, the following is the structure of butyric or butanoic anhydride:

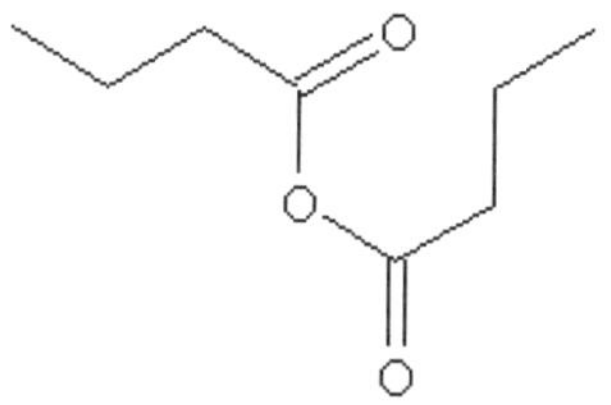

butanoic anhydride

Cyclic anhydrides are named for the dicarboxylic acids from which they are derived. For example, the following is maleic anhydride, derived from maleic acid, HOOCCH=CHCOOH.

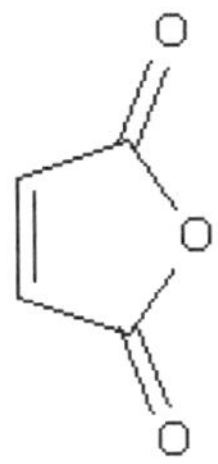

maleic anhydride

Esters are named as alkyl alkanoates. In their names, the alkyl group is written followed by the name of the acyl component, RCO–. For example, the following is *t*-butylbenzoate:

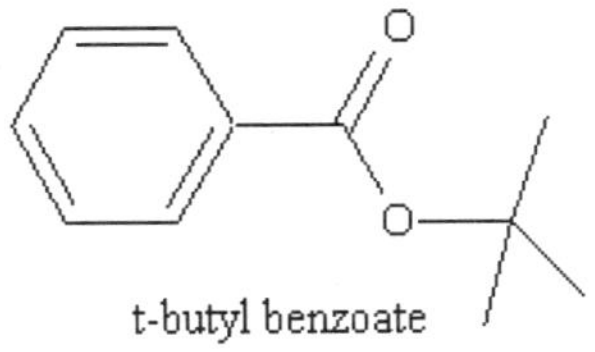

t-butyl benzoate

Cyclic esters are called lactones. The names of amides are also derived from the acids from which they were derived. Thus, *ic acid* or *oic acid* is replaced with *amide*. For example, $HCONH_2$, is called either formamide or methanamide.

The stability of acid derivatives depends on the group bonded to the acyl group. An electron-releasing group such the $-NH_2$ group from amides can stabilize the carbonyl C atom, and an electron-withdrawing group such as a halide atom tends to destabilize the carbonyl C atom. Compounds that are more stabilized are less reactive and those that are less stabilized are more reactive. The order of reactivity of the acid derivatives to hydrolysis is as follows:

$$RCONH_2 < RCOOR´ < RCOOCOR´ < RCOCl$$

Acyl chlorides, most commonly called acid chlorides, can be prepared from carboxylic acids using thionyl chloride, $SOCl_2$.

$$RCOOH + SOCl_2 \rightarrow RCOCl$$

Because acid chlorides are reactive, they react with many compounds. Many of these have been discussed previously. For example, they are used in Friedel-Crafts acylation reactions to form aromatic ketones. They can be reduced to aldehydes or can react with lithium diorganocuprates to form ketones. Acid chlorides also react with organocadmium compounds and produce ketones. They react with alcohols to produce esters.

$$RCOCl + R´OH \rightarrow RCOOR´ + HCl$$

Acid chlorides react with carboxylic acids and produce anhydrides.

$$RCOCl + R´COOH \rightarrow RCOOCOR´ + HCl$$

They also react with ammonia and primary amines in the presence of a base to produce amides.

$$RCOCl + R'NH_2 + OH^- \rightarrow RCONHR' + H_2O + Cl^-$$

Carboxylic acid anhydrides are often prepared from acid chlorides and carboxylate salts.

$$RCOCl + R'COO^- \rightarrow RCOOCOR'$$

Cyclic anhydrides can be formed by heating carboxylic acids in the presence of a dehydrating agent. For example, maleic anhydride is prepared by heating maleic acid in an inert solvent. Acid anhydrides are readily converted to other acid derivatives. They are hydrolyzed to acids.

$$RCOOCOR' + H_2O \rightarrow RCOOH + R'COOH$$

When anhydrides react with alcohols, they produce esters and carboxylic acids.

$$RCOOCOR + R'OH \rightarrow RCOOR' + RCOOH$$

They also react with amines to produce amides and acids.

$$RCOOCOR + R'NH_2 \rightarrow RCONHR' + RCOOH$$

Esters can be prepared in many ways. Acids react with alcohols to form esters.

$$RCOOH + R'OH \rightarrow RCOOR' + H_2O$$

Acid chlorides and acid anhydrides undergo similar reactions but release HCl and RCOOH, respectively. Acids can react with diazomethane, CH_2N_2, and form methyl esters.

$$RCOOH + CH_2N_2 \rightarrow RCOOCH_3 + N_2$$

Esters also undergo many reactions. They hydrolyze to the acids and alcohols from which they were formed. Esters can react with alcohols and produce new esters (transesterification).

$$RCOOR' + R''OH \rightarrow RCOOR'' + R'OH$$

Esters react with amines and produce amides and alcohols.

$$RCOOR' + R'NH_2 \rightarrow RCONHR' + R'OH$$

They are reduced by $LiAlH_4$ to alcohols.

$$RCOOR' + 1.\ LiAlH_4/2.\ H_2O \rightarrow RCH_2OH + R'OH$$

Esters react with Grignard reagents and form tertiary alcohols.

$$RCOOR' \xrightarrow[\text{2. H}_2\text{O}]{\text{1. 2R''MgX}} RR''_2COH + R'OH$$

Five- and six-membered lactones can be produced from hydroxy acids.

Amides can be synthesized by reacting amines with carboxylic acids, acid chlorides, anhydrides, or esters. Primary amides can also be formed by acid or base hydrolysis of a nitrile.

$$RCN + H_2O \rightarrow RCONH_2$$

Amides hydrolyze to acids and amines. They are reduced by LiAlH4 to amines or dehydrated to nitriles.

$$RCONH_2 \xrightarrow[\text{2. H}_2\text{O}]{\text{1. LiAlH}_4} RCH_2NH_2$$

$$RCONH_2 + POCl_3 \rightarrow RCN$$

Amides undergo the Hofmann rearrangement and produce primary amines with one less C atom than the original amides.

$$RCONH_2 + Br_2/OH^- \rightarrow RNH_2 + CO_3^{2-}$$

When γ- and δ-amino acids are heated they produce lactams, cyclic amides.

Test Yourself

1. List the names and general formulas of four groups of carboxylic acid derivatives.

2. Draw the structure of benzoic anhydride.

3. Draw the structure of 4-cyclopentyl-3-butenoyl chloride.

4. Write the name of the following compound.

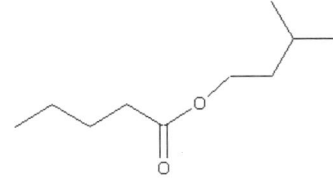

5. Write the name of the following compound.

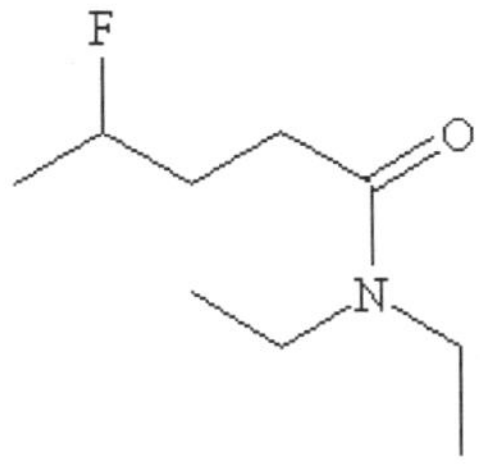

6. What general class of acid derivatives result when RCOCl reacts with each of the following?

   a. R′NH₂

   b. R′COO⁻ M⁺

   c. R′OH

7. Write a mechanism that shows the hydrolysis of acetyl chloride to acetic acid.

8. Write an equation that shows how the following compound is synthesized.

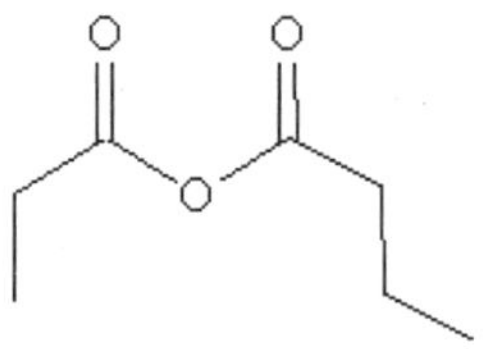

9. Starting with the following compound show how the anhydride of cyclohexane-1,2-dicarboxylic acid could be synthesized.

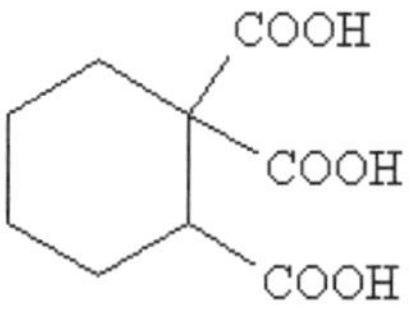

10. What product forms when one mole of the following molecule reacts with one mole of benzene in the presence of AlCl₃?

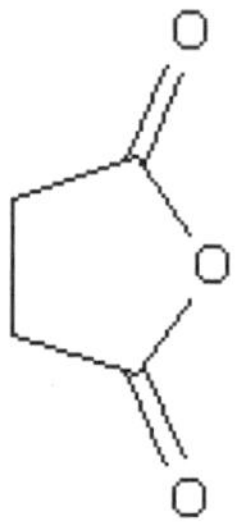

11. Cyclopentane carboxylic acid reacts with ethanol and an acid catalyst, producing Compound *A*. *A* is then treated with LiAlH₄. What product(s) result from this reaction?

12. Consider the following lactone.

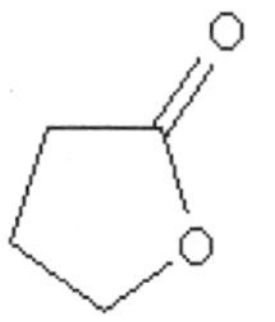

This lactone is mixed for a period of time with an acidic aqueous solution that has ^{18}O instead of ^{16}O in the water molecules. After a period of time, the lactone is isolated. How will the lactone change after being recovered?

13. Write the first two steps of the mechanism of the base-catalyzed hydrolysis of ethyl acetate.

14. What is the product of the reaction of cyclopentylamine and methyl benzoate.

15. How can phthalimide be produced from phthalic acid?

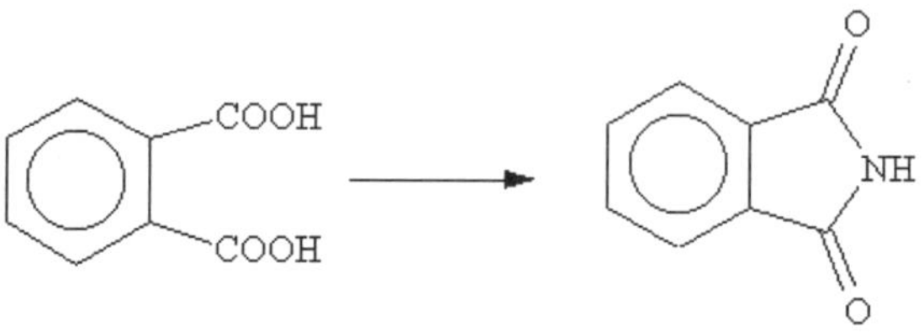

phthalic acid   phthalimide

16. Explain why the rate that methyl acetate undergoes base-catalyzed hydrolysis is significantly faster than the base-catalyzed hydrolysis of *t*-butyl acetate.

17. Draw the structure of the product that results when *p*-chlorobenzamide is treated with bromine in aqueous base.

18. Using only ethanol and all necessary inorganic reagents show how lactic acid, CH₃CH(OH)COOH, can be synthesized.

19. Starting with 4-phenyl-1-butanol show how α-tetralone could be synthesized.

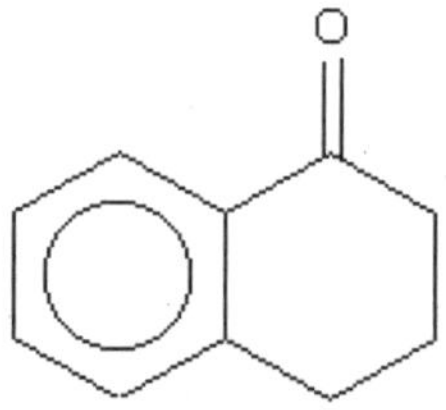

α-tetralone

20. Which of the following most readily undergo hydrolysis?

    a. amides

    b. esters

    c. anhydrides

    d. acyl chlorides

    e. all react at the same rate

    f. none of these

21. Which of the following reactions does **not** take place?

    a. $CH_3COBr + CH_3OH \rightarrow CH_3COOCH_3 + HBr$

    b. $CH_3COOH + CH_3OH \rightarrow CH_3COOCH_3 + H_2O$

    c. $CH_3COOH + NH_3 \rightarrow CH_3CONH_2 + H_2O$

    d. $CH_3COOCOCH_3 + NaOH \rightarrow CH_3COOH + CH_3COO^- Na^+$

    e. all of these reactions occur

22. What compound forms when 4-aminobutanoic acid is heated?

    a. γ-butyrolactam

    b. γ-butyrolactone

    c. succinimide

    d. none of these

23. Which of the following is least soluble in water?

    a. ethyl acetate

    b. dimethylformamide

    c. dimethylacetamide

    d. acetonitrile

24. Which of the following molecules is least stable?

    a. δ-valerolactone

    b. β-propiolactone

    c. δ-valerolactam

    d. γ-butyrolactam

25. Which of the following converts hexanamide to hexanenitrile?

    a. 1. $LiAlH_4$, 2. H_2O

    b. Br_2, OH^-

    c. $NaCN$, H^+

    d. none of these

26. Which of the following is a carbamate?

    a. $CH_3CH_2CON(CH_3)_2$

    b. $CH_3CH_2NHCOOCH_3$

    c. $CH_3CH_2OCOOCH_2CH_3$

    d. none of these

27. What product initially results when butyric acid reacts with ethylamine?

    a. ethylammonium butyrate

    b. *N,N*-diethylbutanamide

    c. *N*-ethylbutanamide

    d. none of these

28. Which of the following is **not** correct about the spectra of *all* amides?

    a. In the IR spectrum, 1° and 2° amides have strong carbonyl stretch peaks in the range of 1640 to 1680 cm^{-1}.

    b. In the IR spectrum, amides have N–H stretching absorptions in the range of 3200 to 3500 cm^{-1}.

    c. In the ^1H NMR spectrum, 1° and 2° amides have a broad peak in the range of 5 to 8 ppm (δ).

    d. In the ^{13}C NMR spectrum, amides have a peak in the range of 160 to 180 ppm (δ).

29. What reagent could be used to make the following conversion?

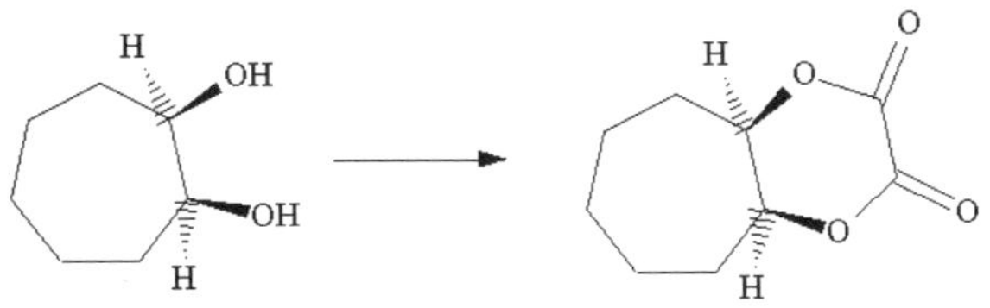

    a. $CH_3COOCOCH_3$

    b. CH_3COCl

    c. $CH_3CH_2CH_2COCl$

    d. $ClOCCH_2CH_2COCl$

    e. $ClOCCH_2COCl$

    f. none of these

30. Which is the product of the following reaction?

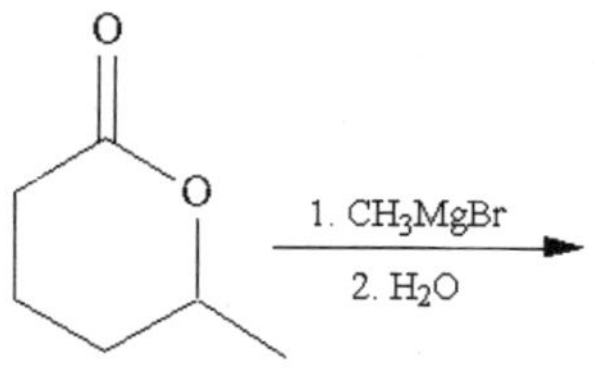

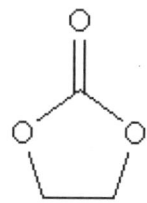

1. CH₃MgBr
2. H₂O

  a. 2,6-heptanediol

  b. 2-methyl-2,6-heptanediol

  c. 2-methyl-2-heptanol

  d. 2-methyl-2,5-heptanediol

  e. none of these

31. Which of the following esters most readily undergoes alkaline hydrolysis?

  a. methyl *p*-methoxybenzoate

  b. methyl *p*-methylbenzoate

  c. methyl *p*-nitrobenzoate

  d. methyl *p*-bromobenzoate

32. Consider the compound *N*-methylsuccinimide.

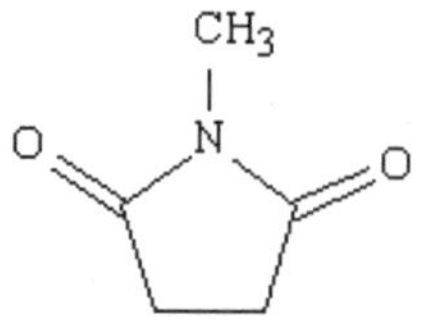

One mole of this compound is treated with two moles of NaOH and then the product is hydrolyzed with acid. What is the structure of the product?

  a. HOOCCH₂CH₂CONH₂

  b. H₂NCOCH₂CH₂CONH₂

  c. HOOCCH₂CH₂COOH

  d. CH₃CH₂CH₂COOH

  e. none of these

33. What is the product of the reaction of pentamide and P₂O₅?

  a. pentanenitrile

  b. pentanoic acid

  c. 1-pentanol

  d. no reaction

34. Which of the following will react in the presence of a Lewis acid catalyst to produce the following?

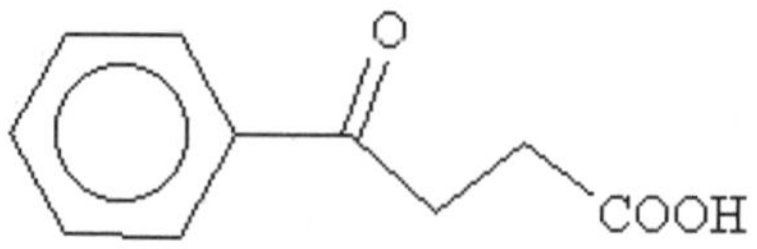

  a. benzene and maleic anhydride

  b. benzene and succinic anhydride

  c. benzyl alcohol and acetic anhydride

  d. benzene and phthalic anhydride

  e. none of these

35. What two reactants produce the following compound?

  a. ethylene glycol and phosgene (COCl₂)

  b. propylene glycol and phosgene (COCl₂)

  c. ethylene glycol and acetyl chloride

  d. propylene glycol and acetyl chloride

  e. none of these

 **Check Yourself**

1. 1. Esters, RCOOR´
    2. Acid or acyl halides, RCOX
    3. Acid anhydrides, RCOOCOR´
    4. Amides, RCONH₂, RCONHR, and RCONR₂. (**Acid derivatives**)

2. The structure of benzoic anhydride is as follows. (**Acid derivative structures**)

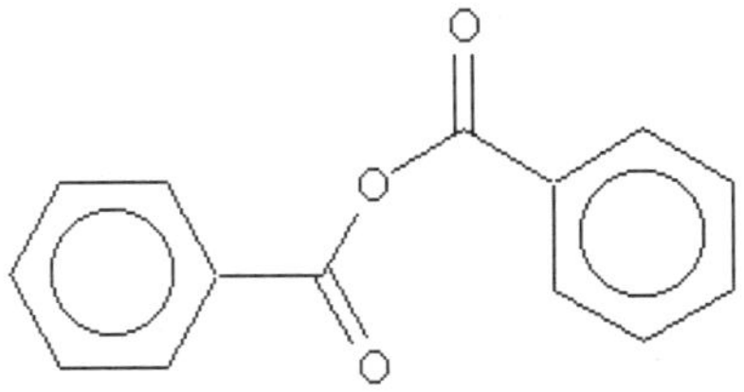

3. The structure of 4-cyclopentyl-3-butenoyl chloride is as follows.

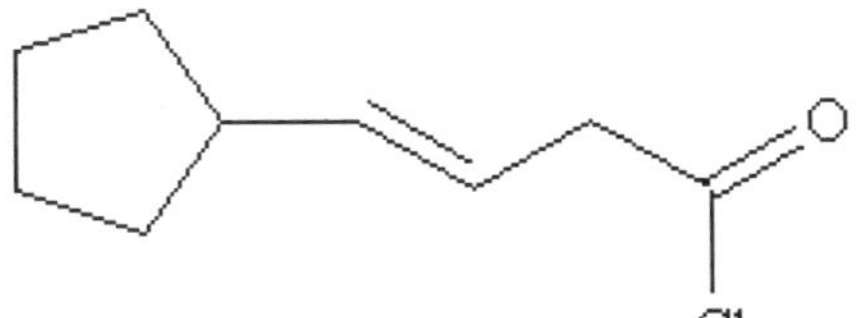

    4-cyclopentyl-3-butenoyl chloride

    (**Acid derivative structures**)

4. The name of this compound is 3-methylbutyl pentanoate, an ester. (**Acid derivative nomenclature**)

5. The name of this compound is *N,N*-diethyl-4-fluoro-pentanamide. (**Acid derivative nomenclature**)

6. a. Amide, RCONHR´
    b. Acid anhydride, RCOOCOR´
    c. Ester, RCOOR´ (**Acid derivatives**)

7. The mechanism for the hydrolysis of acetyl chloride to acetic acid is as follows.

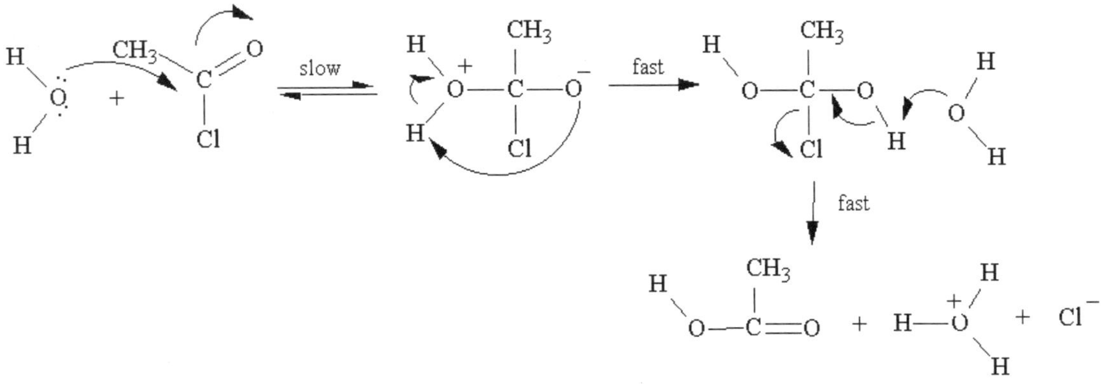

    (**Acid chloride reactions**)

8. This compound is an acid anhydride, which can be prepared as follows from propanoyl chloride and butanoic acid in pyridine.

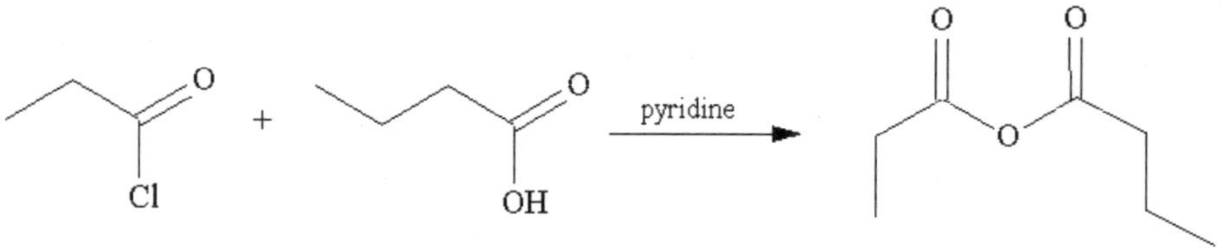

Another way to synthesize this product is to react butanoyl chloride and propanoic acid. **(Anhydride synthesis)**

9. Start by decarboxylating 1,1,2-cyclohexanetricarboxylic acid and then continue heating to drive out water, producing the anhydride.

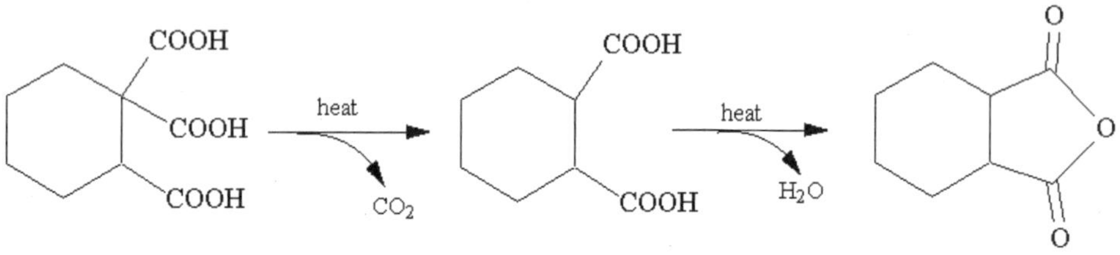

(Anhydride synthesis)

10. When succinic acid reacts with benzene it undergoes a Friedel-Crafts acylation reaction. Only one succinic acid molecule will add to the anhydride because the added substituent group deactivates the benzene ring.

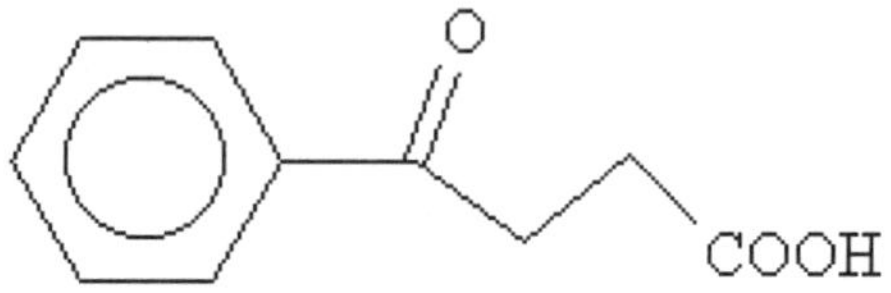

(Acid anhydride reactions)

11. Cyclopentane carboxylic acid reacts with ethanol and an acid catalyst, producing the ethyl ester, which when treated with $LiAlH_4$ produces two alcohols. **(Ester synthesis/reactions)**

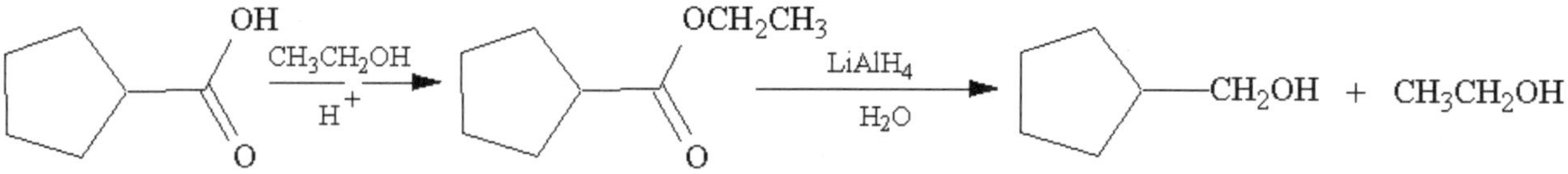

12. This lactone reacts through a nucleophilic addition reaction with the $H_2{}^{18}O$ forming a tetrahedral intermediate. This intermediate changes back to the lactone by eliminating either the ^{16}O-labeled water or the ^{18}O-labeled water; thus, some carbonyl O atoms become labeled with ^{18}O. (**Ester reactions**)

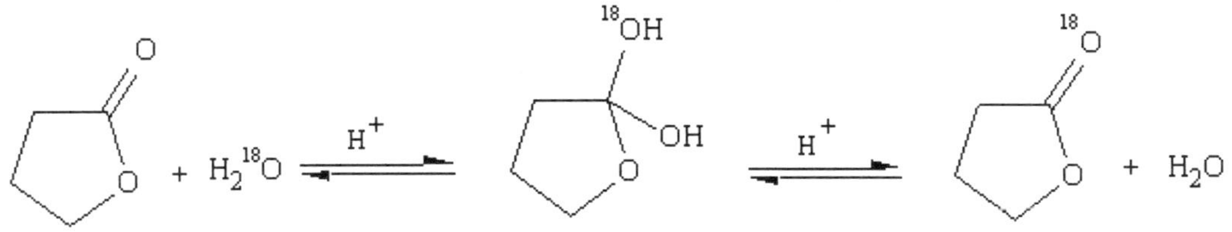

(**Ester reactions**)

13. The first step is the attack on the carbonyl group by the OH^-. This forms the anionic form of the tetrahedral intermediate. In the second step, acetic acid and ethoxide result when the "extra" pair of electrons on the O atom reforms the C–O double bond.

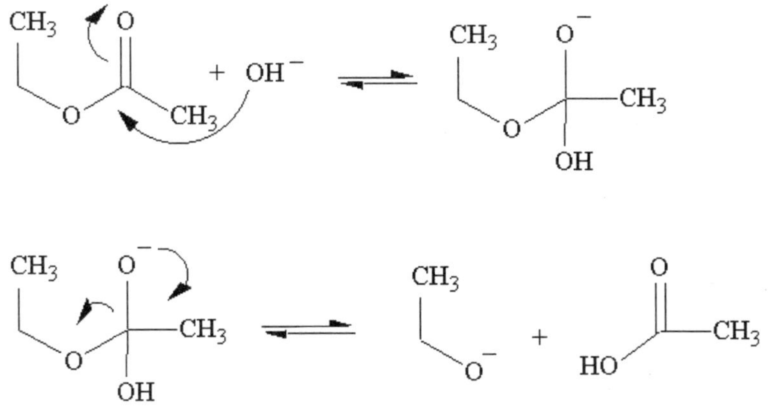

(**Ester reactions**)

14. Esters react with primary amines and produce substituted amides. The reaction of methyl benzoate and cyclopentylamine produces *N*-cyclopentylbenzamide and methanol.

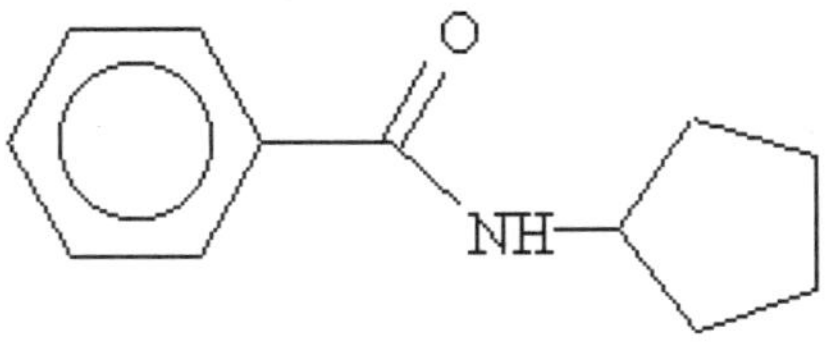

(**Amide formation**)

15. First, dehydrate phthalic acid to phthalic anhydride. Then react phthalic anhydride with two moles of ammonia, producing ammonium phthalamate. Finally, heating ammonium phthalamate drives off ammonia and water, producing phthalimide. A second way is to first react with two moles of $SOCl_2$, forming the diacid chloride. Then react this compound with ammonia followed by aqueous base. **(Imide synthesis)**

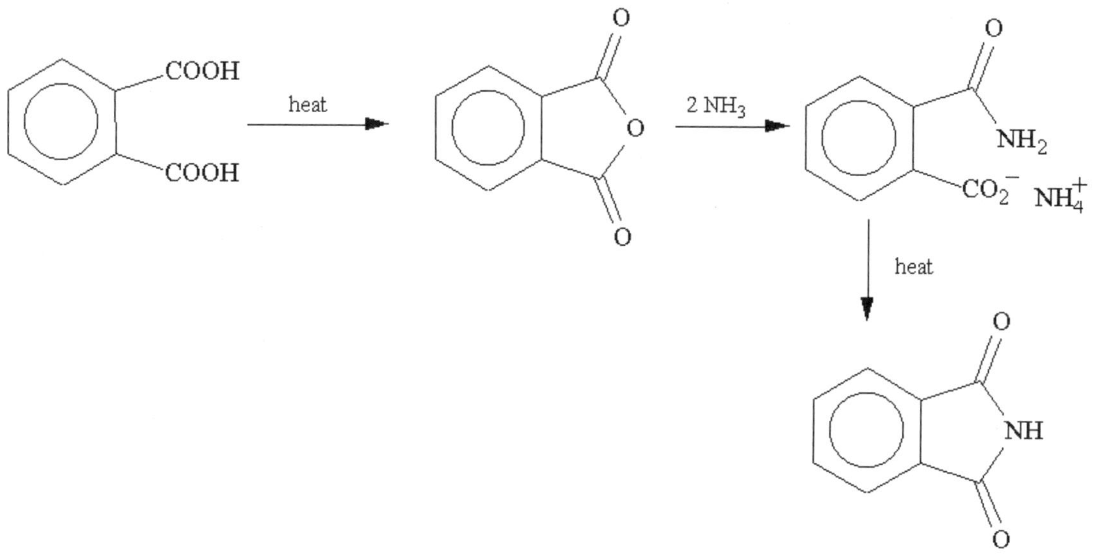

16. In the base-catalyzed hydrolysis of an ester, a tetrahedral intermediate forms. More stable intermediates form faster than less stable ones. The intermediate from methyl acetate is much less crowded and thus more stable than the crowded *t*-butyl tetrahedral intermediate. The *t*-butyl intermediate forms slower because it is less stable. **(Hydrolysis of esters)**

17. When an amide reacts with Br_2 in base it undergoes the Hofmann rearrangement. The product of this reaction is *p*-chloroaniline.

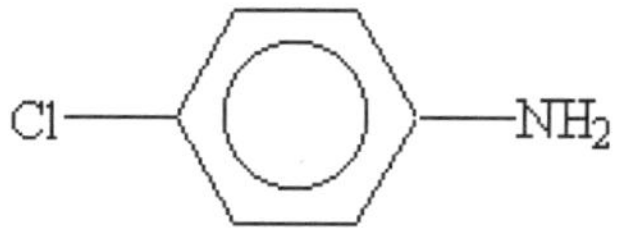

(Hofmann rearrangement)

18. Because ethanol has two C atoms and lactic acid has three C atoms, the length of the chain must be increased by one. This is accomplished as follows.

$CH_3CH_2OH + PCl_3 \rightarrow CH_3CH_2Cl + CN^- \rightarrow CH_3CH_2CN + H^+(aq)/\text{heat} \rightarrow CH_3CH_2COOH$

To add the OH group to the second C atom first substitute a halogen atom on this C atom and then do a nucleophilic displacement to add the OH group. Since OH^- is used for the nucleophilic substitution, acid must be added to convert the carboxylate salt to the carboxylic acid.

$CH_3CH_2COOH + Br_2/PCl_3 \rightarrow CH_3CHBrCOOH + OH^- \rightarrow CH_3CH(OH)COO^-$

$CH_3CH(OH)COO^- + H^+ \rightarrow CH_3CH(OH)COOH$

(Acid synthesis)

19. First, oxidize the alcohol to an acid using chromic acid. Then convert the acid to an acyl chloride, using $SOCl_2$. Finally, do a Friedel-Crafts acylation using $AlCl_3$. **(Acid reactions)**

20. d. Acyl chlorides are hydrolyzed most readily because the Cl atom does little to stabilize the carbonyl C atom. All of the other acid derivatives have a larger degree of stabilization of the carbonyl C atom. **(Acid derivative reactions)**

21. e. All of these reactions occur. The reaction of CH_3COOH and NH_3 produces the ammonium salt with acetic acid, but if heated will produce acetamide. **(Acid derivative reactions)**

22. a. γ-Butyrolactam forms when 4-aminobutanoic acid is heated.

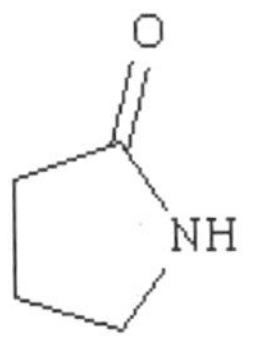

(Lactam synthesis)

23. a. Ethyl acetate is only partially miscible with water. The others are totally miscible with water. **(Properties of acid derivatives)**

24. b. β-Propiolactone is the most unstable molecule because it has the greatest strain within its four-membered ring. All of the others have five or more atoms in the ring. **(Properties of acid derivatives)**

25. d. None of these will convert hexamide to hexanenitrile. To accomplish this, a strong dehydrating agent such as P_2O_5 or $POCl_3$ is used. **(Amide reactions)**

26. b. $CH_3CH_2NHCOOCH_3$ is a carbamate. They have the general formula of RNHCOOR′. **(Carbamates)**

27. a. Ethylammonium butyrate forms first when ethylamine and butyric acid react. When ethylammonium butyrate is heated for a period of time, it then forms the *N*-ethylbutanamide. **(Amide synthesis)**

28. b. In the IR spectrum, only primary and secondary amides have N–H stretching absorptions in the range of 3200 to 3500 cm⁻1. **(Amide spectra)**

29. f. The reagent need for this conversion is as follows.

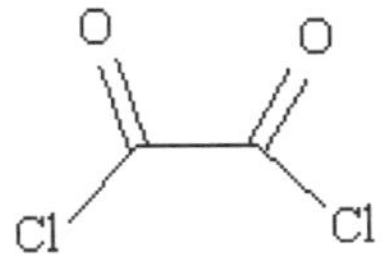

(Ester synthesis)

30. b. 2-Methyl-2,6-heptanediol is the product of this reaction. Grignard reagents react with esters and produce tertiary alcohols. **(Ester reactions)**

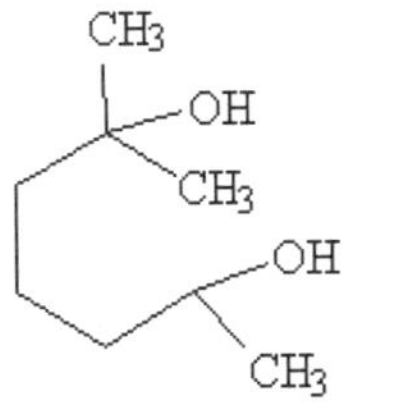

2-methyl-2,6-heptanediol

31. c. Methyl *p*-nitrobenzoate will most readily undergo alkaline hydrolysis because electron-withdrawing groups help disperse the developing negative charge in the transition state and makes the carbonyl C atom more electrophilic. (**Ester reactions**)

32. c. $HOOCCH_2CH_2COOH$ is the product of the reaction because the hydroxide cleaves both amide bonds. This produces the sodium succinate, which hydrolyzes in acid to succinic acid. (**Amide reactions**)

33. a. A strong dehydrating agent such as P_2O_5 or $POCl_3$ converts pentamide to pentanenitrile, $CH_3CH_2CH_2CH_2CN$. (**Amide reactions**)

34. b. Benzene and succinic anhydride react to form the given compound, 4-oxo-4-phenylbutanoic acid. (**Anhydride reactions**)

35. a. Ethylene glycol, $HOCH_2CH_2OH$, and phosgene, $ClCOCl$, react to produce this compound. (**Carbonate esters**)

Grade Yourself

Circle the number of questions you missed, then fill in the total incorrect for each topic. If you answered more than three questions incorrectly, you need to focus on that topic. (If a topic has less than three questions and you had at least one wrong, we suggest you study that topic also. Read your textbook, a review book, or ask your teacher for help.)

Subject: Carboxylic Acid Derivatives

Topic	Question Numbers	Number Incorrect
Acid derivatives	1, 6	
Acid derivative structures	2, 3	
Acid derivative nomenclature	4, 5	
Acid chloride reactions	7	
Anhydride synthesis	8, 9	
Acid anhydride reactions	10	
Ester synthesis/reactions	11, 12, 13, 29, 30, 31	
Amide formation	14	
Imide synthesis	15	
Hydrolysis of esters	16	
Hofmann rearrangement	17	
Acid synthesis	18	
Acid reactions	19	
Acid derivative reactions	20, 21	
Lactam synthesis	22	
Properties of acid derivatives	23, 24	
Amide reactions	25, 32, 33	
Carbamates	26	
Amide synthesis	27	
Amide spectra	28	
Anhydride reactions	34	
Carbonate esters	35	

Reactions of β-*Dicarbonyl Compounds*

19

Brief Yourself

One of the most important groups of ester enolates is the β-keto esters, RCOCH₂COOR′. These and related compounds are important in synthetic organic chemistry. The H atoms α to both carbonyl groups are acidic because of the resonance delocalization of the anion. For example, resonance of the acetoacetate anion is as follows.

One of the most important ester enolate reactions is the Claisen condensation. In this reaction, two esters with α H atoms condense to a β-keto ester in the presence of a strong base.

$$2\ RCH_2COOR' + CH_3CH_2O^- \rightarrow RCH_2COCHRCOOR'$$

The Claisen condensation may also be run with two different esters. To obtain reasonable yields one of the esters must not have α H atoms.

$$RCH_2COOR'' + R'COOR'' \ (CH_3CH_2O^-) \rightarrow R'COCHRCOOR''$$

Mixed Claisen condensations often use formate esters, carbonate esters, or benzoate esters. Esters of some dicarboxylic acids undergo an intramolecular Claisen condensation. This reaction, called the Dieckmann condensation, gives reasonable yields when the reactants produce either five- or six-membered rings.

Another important ester enolate reaction is the acetoacetic ester synthesis. In this reaction ethyl acetoacetate first reacts with a base followed by an alkyl halide. The alkyl group from the halide substitutes for an α H atom.

$$CH_3COCH_2COOCH_2CH_3 + RX\ (CH_3CH_2O^-) \rightarrow CH_3COCHRCOOCH_2CH_3$$

When this compound is hydrolyzed and decarboxylated, an α-alkylated acetone derivative, CH_3COCH_2–R, forms.

$$CH_3COCHRCOOCH_2CH_3 + 1.\ H^+/2.\ heat \rightarrow CH_3COCH_2\text{–}R$$

The malonic ester synthesis is a reaction similar to the acetoacetic ester pathway. In this reaction, an alkyl halide reacts with malonic ester, diethyl malonate, producing an α-substituted diethyl malonate, which after hydrolysis and decarboxylation produces an α-substituted acetic acid.

$$CH_2(CO_2CH_2CH_3)_2 + RX\ (CH_3CH_2O^-) \rightarrow R\text{–}CH(CO_2CH_2CH_3)_2$$

$$R\text{–}CH(CO_2CH_2CH_3)_2 + 1.\ H^+/2.\ heat \rightarrow RCH_2COOH$$

Because malonic ester has two α H atoms, two successive alkylation steps can replace both H atoms, producing an acetic acid derivative with two alkyl groups, RR′CHCOOH.

Enolate anions, and other stabilized anions tend to undergo conjugate addition to α,β-unsaturated compounds. This reaction is the Michael reaction or Michael addition. One example of a Michael reaction is the reaction of diethyl malonate, a Michael donor, with an α,β-unsaturated carbonyl compound, a Michael acceptor.

$$RCOCH{=}CH_2 + CH_2(CO_2CH_2CH_3)_2 \rightarrow RCOCH_2CH_2CH(CO_2CH_2CH_3)_2$$

Many groups of compounds are Michael donors, including β-diketones, β-keto esters, β-keto nitriles, and enamines. Many groups of compounds are Michael acceptors, including conjugated aldehydes, ketones, esters, amides, and nitriles.

Diethyl malonate or acetoacetic ester reacts with aldehydes and ketones in a manner similar to mixed aldol condensations. This means that the α C atom of the diethyl malonate bonds to the carbonyl C atom of the aldehyde or ketone. This is followed by the elimination of water.

$$RCHO + CH_2(CO_2CH_2CH_3)_2 \rightarrow RCH{=}C(CO_2CH_2CH_3)_2 + H_2O$$

A small amount of an amine, such as piperidine, is added to catalyze this reaction. This reaction is called the Knoevenagel condensation.

Another ester enolate reaction is the Reformatsky reaction in which an enolate ion, produced from the reduction of an α-bromo ester by Zn, reacts with an aldehyde or ketone, producing the ester of a β-hydroxy acid.

$$RCHO + R′CHBrCOOCH_2CH_3 + 1.\ Zn/C_6H_6,\ 2.\ H^+ \rightarrow RCH(OH)CHR′COOCH_2CH_3$$

Test Yourself

1. Write the three principal resonance structures for the acetoacetic ester (ethyl acetoacetate) enolate ion.

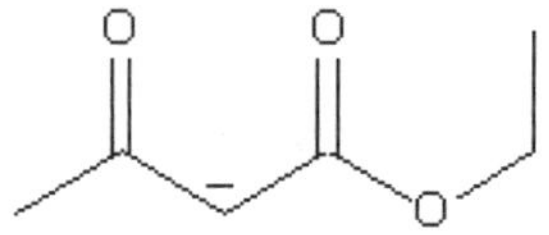

2. a. What is the product of the condensation reaction of two moles of ethyl propanoate with sodium ethoxide followed by acid hydrolysis?

   b. What general type of reaction is this?

3. Draw the structure of the product that results from the Dieckmann condensation of diethyl heptanedioate in base followed by acid hydrolysis.

4. Draw the structure of the product of the mixed Claisen condensation of ethyl benzoate and ethyl phenylacetate, $C_6H_5CH_2COOCH_2CH_3$.

5. Using a Claisen condensation, what two reactants could be used to produce the ethyl (2-oxocyclopentane)carboxylate?

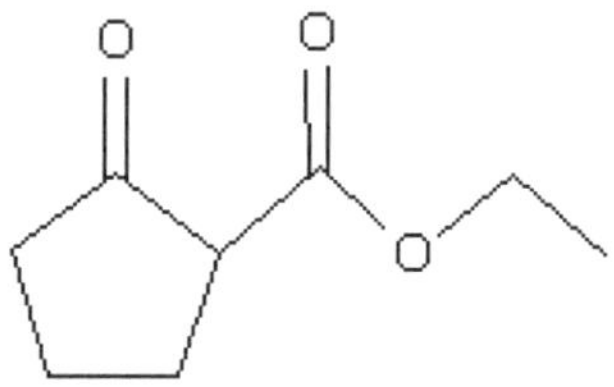

   ethyl (2-oxocyclopentane)carboxylate

6. What product forms if ethyl (2-oxocyclopentane)carboxylate, the compound in Problem 5, is first reacted with aqueous base, then treated with acid and heated?

7. 1-Bromo-3-cyclopentylpropane and ethyl acetoacetate are first reacted in a strong basic solution. The product of this reaction is hydrolyzed in a strong basic solution and then converted to an acid which is decarboxylated. What is the product of this synthesis?

8. What reactants are needed to perform an acetoacetic ester synthesis of the following compound?

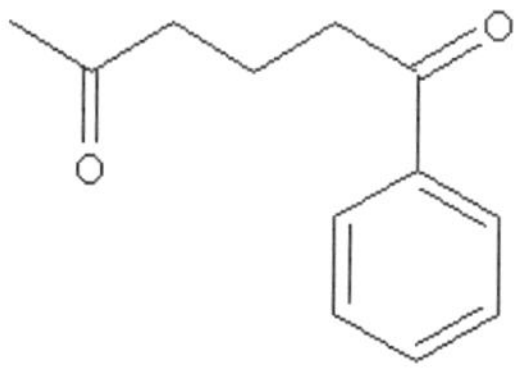

9. Explain the difference in the products obtained from the acetoacetic ester and the malonic ester syntheses.

10. a. Write the equation for the reaction of sodium ethoxide and diethyl malonate (also called malonic ester).

    b. Explain the acidity of malonic ester.

11. How is 2-ethylhexanoic acid prepared by a malonic ester synthesis?

12. What reactant is needed to synthesize cyclopentanecarboxylic acid in a malonic ester synthesis?

13. a. What reactants are typically used in the Michael reaction?

    b. What general classes of compounds are produced from Michael reactions?

14. What is the product of the following reaction?

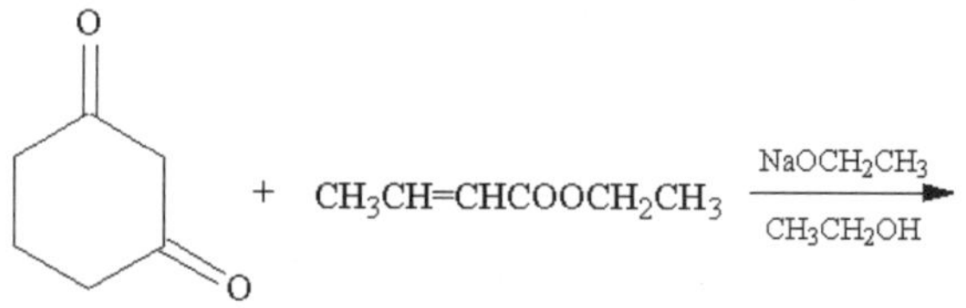

$$+ \quad CH_3CH{=}CHCOOCH_2CH_3 \xrightarrow[CH_3CH_2OH]{NaOCH_2CH_3}$$

15. What product results from the Robinson annulation using the following compounds?

16. What two reactants are required to produce the following compound using a Robinson annulation?

17. What reactants are needed to produce the following compound using the Reformatsky reaction?

ethyl 3-hydroxy-2,3-dimethylpentanoate

18. 6,6-Dimethyl-2-cyclohexenone reacts with diethyl malonate in the presence of sodium ethoxide and ethanol and produces Compound A. A is then hydrolyzed and then heated, producing compound B. Draw the structures of A and B.

19. a. What is the product of the Knoevenagel condensation of pentanal and diethylmalonate?

    b. What conditions are required for this reaction?

20. Which of the following **cannot** undergo a Claisen condensation reaction with another molecule of itself?

    a. $C_6H_5CH_2COOCH_3$
    b. $C_6H_5COOCH_3$
    c. $CH_3CH_2COOC_6H_5$
    d. $CH_3CH_2CH_2CH_2COOC_6H_5$

21. Which of the following could **not** be synthesized using the acetoacetic ester synthesis?

    a. methyl phenyl ketone
    b. methyl ethyl ketone
    c. methyl butyl ketone
    d. methyl isopropyl ketone
    e. none of these

22. Which of the following alkyl halides is required to synthesize 6,6-dimethyl-2-heptanone using an acetoacetic ester synthesis?

    a. 1-bromo-3-methylbutane
    b. 1-bromo-3-methylpentane
    c. 1-bromo-3,3-dimethylpentane
    d. 1-bromo-3,3-dimethylbutane
    e. none of these

23. What compound results when *n*-hexylmalonic acid is heated?

    a. 2-octanone
    b. 2-hexanone
    c. octanal
    d. hexanoic acid
    e. no reaction
    f. none of these

24. What two alkyl groups must be added to malonic ester to produce 2-methyldecanoic acid in a malonic ester synthesis?

    a. methyl and *n*-octyl
    b. methyl and *n*-hexyl
    c. methyl and *n*-decyl
    d. ethyl and *n*-decyl
    e. none of these

25. Which of the following synthetic pathways gives the best yield of 1,3-diphenyl-5-hexen-2-one starting with ethyl phenylacetate?

    a. Claisen condensation

    b. Dieckmann condensation

    c. malonic ester synthesis

    d. Michael reaction

    e. none of these

26. Which of the following synthetic pathways gives the best yields of diethyl 2-ethyl-1-butene-1,1-dicarboxylate starting with 3-pentanone and any other appropriate reagents?

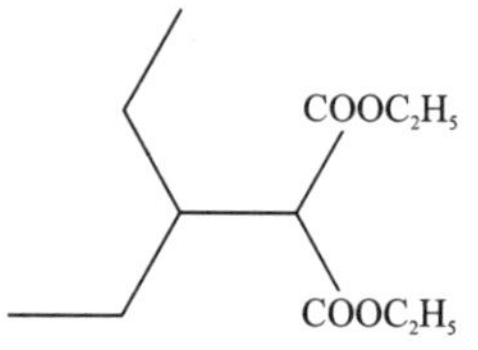

    diethy 2-ethyl-1-butene-1, 1-dicarboxylate

    a. Claisen condensation

    b. Dieckmann condensation

    c. acetoacetic ester synthesis

    d. malonic ester synthesis

    e. Michael reaction

    f. none of these

27. How many different isomeric enol forms of ethyl acetoacetate (disregarding *E-Z* isomers) can be produced?

    a. zero

    b. one

    c. two

    d. three

    e. none of these

28. Which of the following most readily undergoes the Dieckmann condensation?

    a. diethyl propanedioate

    b. diethyl butanedioate

    c. diethyl hexanedioate

    d. diethyl decanedioate

    e. none of these

29. Ethyl acetoacetate undergoes a Michael addition with phenyl vinyl ketone in the presence of sodium ethoxide in ethanol. What product results if the product of this Michael addition undergoes saponification and decarboxylation?

    a. 1-phenyl-1,5-hexanedione

    b. 2-phenyl-1,5-hexanedione

    c. ethyl 2-acetyl-5-oxo-5-phenylpentanoate

    d. none of these

30. What reactants produce the following compound using a Claisen condensation?

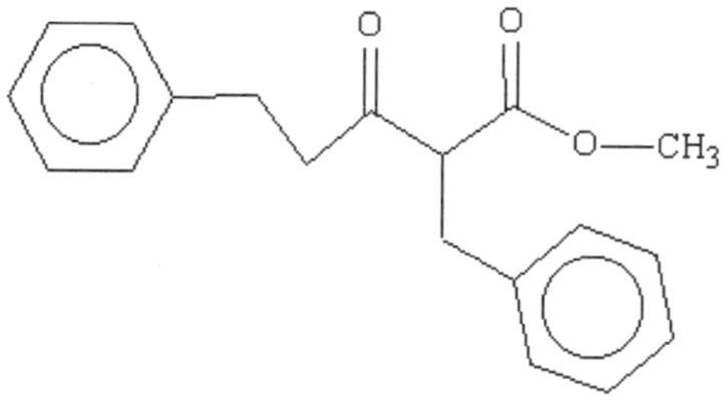

    a. 2 C_6H_5–CH_2COOCH_3

    b. 2 C_6H_5–$CH_2CH_2COOCH_3$

    c. C_6H_5–CH_2COOCH_3 + C_6H_5–$CH_2CH_2COOCH_3$

    d. 2 C_6H_5–$COOCH_2CH_3$

    e. none of these

Check Yourself

1. The three resonance structures of acetoacetic ester enolate ion are represented as follows.

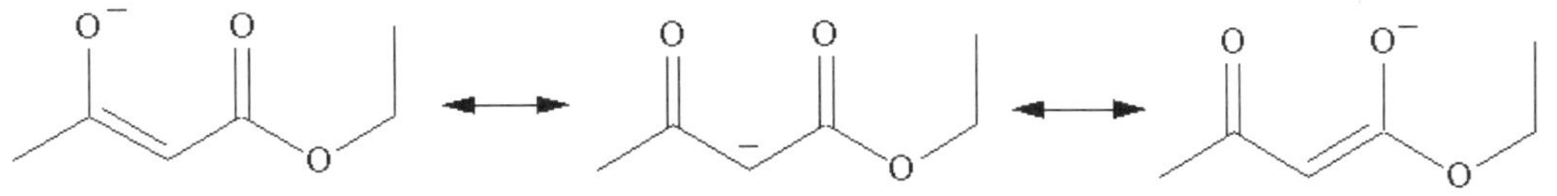

(β-Keto acids)

2. a. When two molecules of ethyl propanoate react in a strong basic solution they produce ethyl 2-methyl-3-oxopentanoate, $CH_3CH_2COCH(CH_3)COOCH_2CH_3$.
   b. This is a Claisen condensation reaction. **(Claisen condensation)**

3. The Dieckmann condensation is an intramolecular Claisen condensation. Thus, the α H atoms from one ester bond to the carbonyl group of the other ester, eliminating a molecule of ethanol and producing the following compound after hydrolysis.

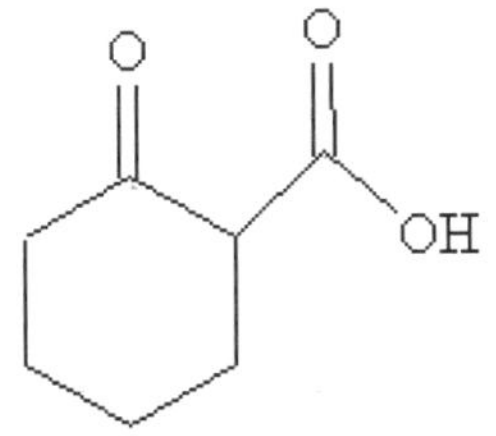

(Dieckmann condensation)

4. In this reaction the α C atom of $C_6H_5CH_2COOCH_2CH_3$ bonds to the carbonyl C atom of ethyl benzoate. The product of this reaction is as follows.

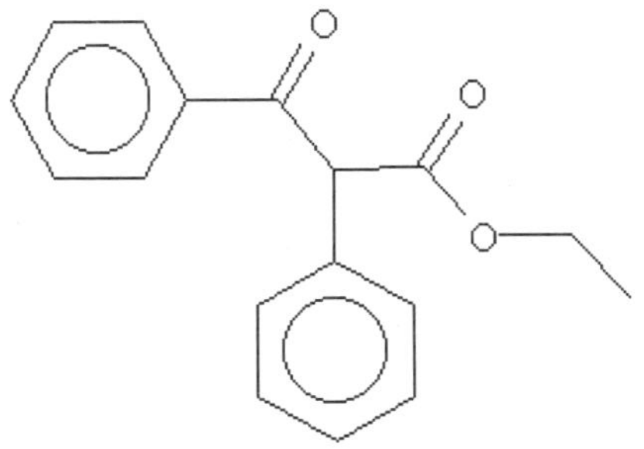

(Mixed Claisen condensation)

5. To synthesize the desired compound, react cyclopentanone and diethyl carbonate in $NaOC_2H_5/CH_3CH_2OH$. **(Mixed Claisen condensation)**

6. Treating ethyl (2-oxocyclopentane)carboxylate with base hydrolyzes the ester and produces the carboxylate salt. It is then converted to the carboxylic acid, which upon heating decarboxylates yielding cyclopentanone. **(Decarboxylation of β-keto esters)**

7. This is an acetoacetic ester synthesis in which the salt of acetoacetic ester does a nucleophilic substitution for the Br atom in 1-bromo-3-cyclopentylpropane producing a 2-alkyl derivative of ethyl acetoacetate. This product is

then converted to the 2-alkyl derivative of the ethylacetic acid, which decarboxylates to the following ketone.

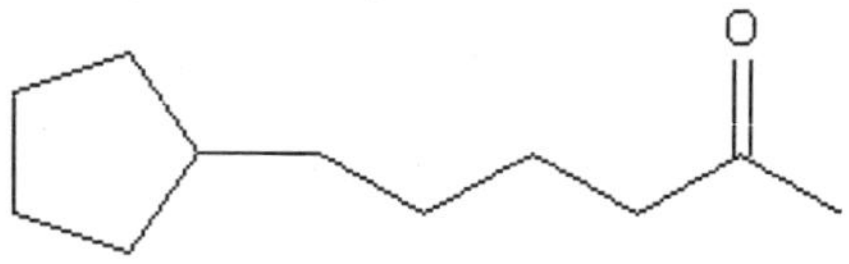

6-cyclopentyl-2-hexanone

(Acetoacetic ester synthesis)

8. Acetoacetic ester syntheses produce α-alkyl substituted acetones, $R-CH_2COCH_3$. The acetone component of the resulting molecule comes from the ethyl acetoacetate and the alkyl component comes from an alkyl halide. Thus, the reactants needed for this reaction are ethyl acetoacetate and 3-bromo-1-phenyl-1-propanone.

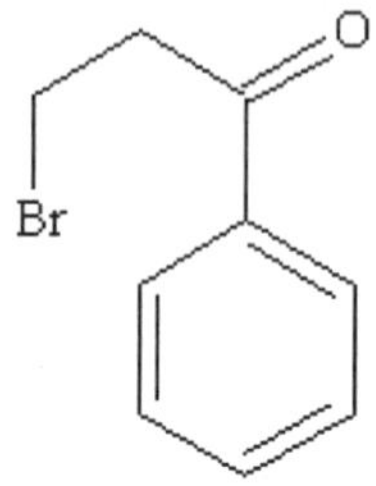

(Acetoacetic ester synthesis)

9. The product of an acetoacetic ester synthesis is an α-alkyl substituted acetone, $R-CH_2COCH_3$. The product of a malonic ester synthesis is an α-alkyl acetic acid, $R-CH_2COOH$. **(Malonic ester synthesis)**

10. a. The equation for sodium ethoxide and malonic ester is as follows.

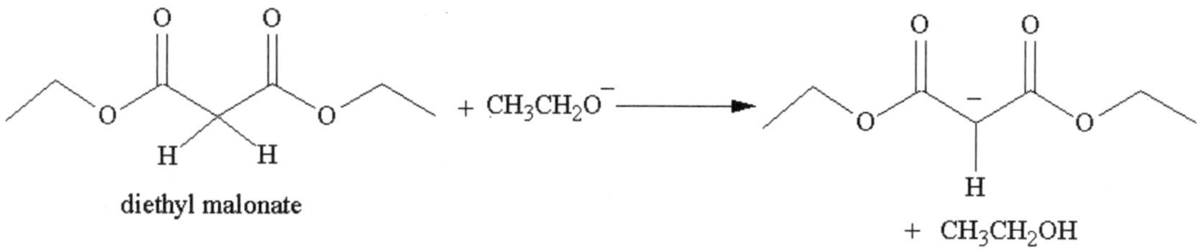

diethyl malonate

$+ CH_3CH_2OH$

b. Malonic ester is acidic because of the resonance stabilization of the conjugate base of the malonic ester. The negative charge is dispersed over the two carbonyl groups and the linking CH_2 group. **(Malonic ester synthesis)**

11. The structure of 2-ethylhexanoic acid is as follows.

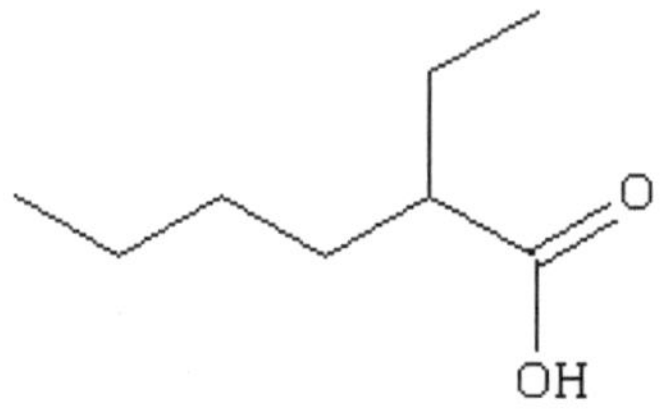

Because the malonic ester synthesis produces alkyl-substituted acetic acid molecules both an ethyl group and *n*-butyl group must be added. Thus, first treat malonic ester with base followed by ethyl bromide and then treat it again with base followed by *n*-butyl bromide (or in reverse order). To obtain the acid, base hydrolysis, addition of acid, and decarboxylation are needed. **(Malonic ester synthesis)**

12. To synthesize cyclopentanecarboxylic acid, malonic ester must react with 1,4-dibromobutane.

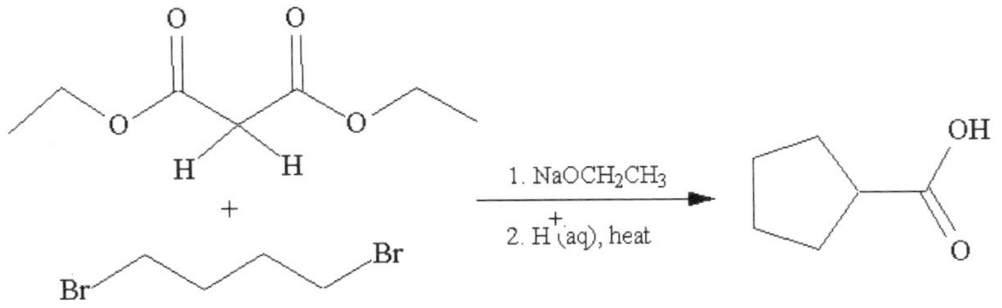

(Malonic ester synthesis)

13. a. The Michael reaction occurs when ,β-unsaturated compounds, Michael acceptors, react with β-keto carbonyl enolates, Michael donors, such as acetoacetic ester or malonic ester.
    b. The Michael reaction is used to synthesize keto acids, keto esters, diketones, and others.
    (Michael addition)

14. This is a Michael addition reaction—a 1,4-addition of a Michael donor to a Michael acceptor.

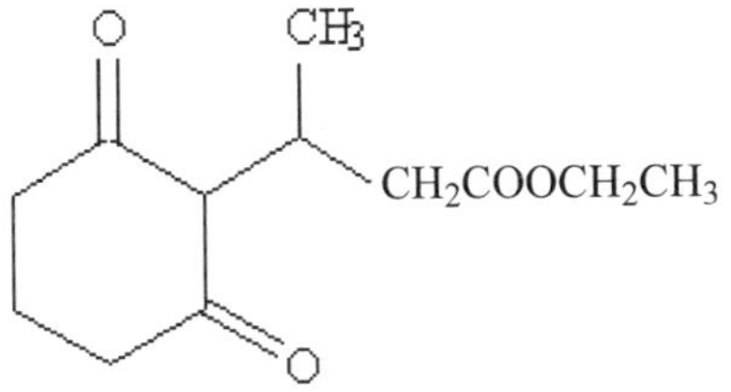

(Michael addition)

15. The product of the Robinson annulation is the result of a Michael reaction followed by an aldol condensation. Thus, the product of this reaction is as follows.

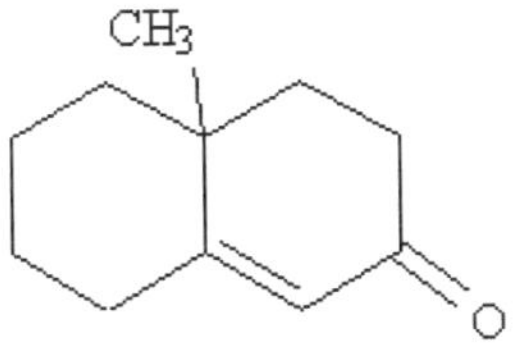

(Robinson annulation)

16. This product can be formed as follows.

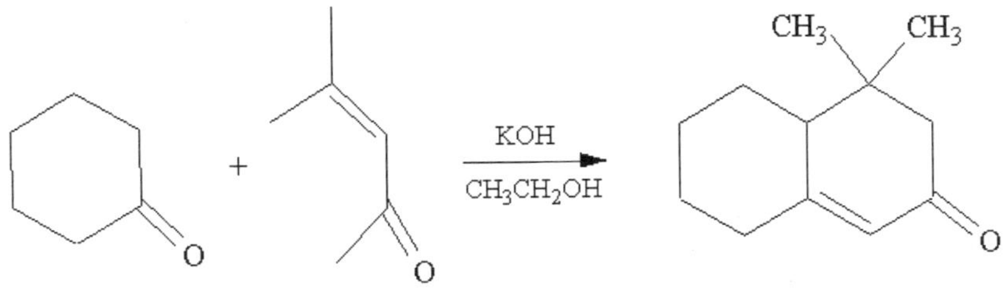

(Robinson annulation)

17. In the Reformatsky reaction, an aldehyde or ketone combines with an α-bromoester, producing the ester of a β-hydroxy acid. Thus, butanone must react with ethyl 2-bromopropanoate to yield the desired product.

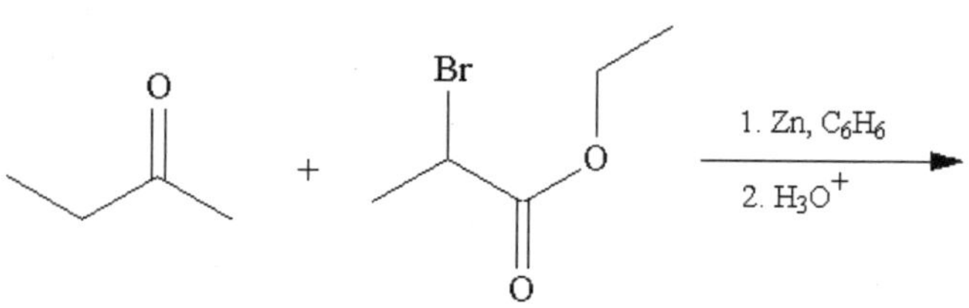

(Reformatsky reaction)

18. Diethylmalonate undergoes a Michael reaction with 6,6-dimethyl-2-cyclohexenone and produces the following compound (*A*).

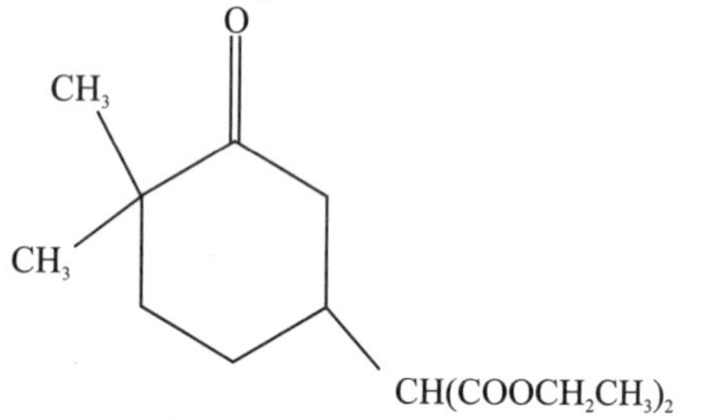

After the addition of acid and then heating, *A* hydrolyzes and decarboxylates to the following compound, *B*.

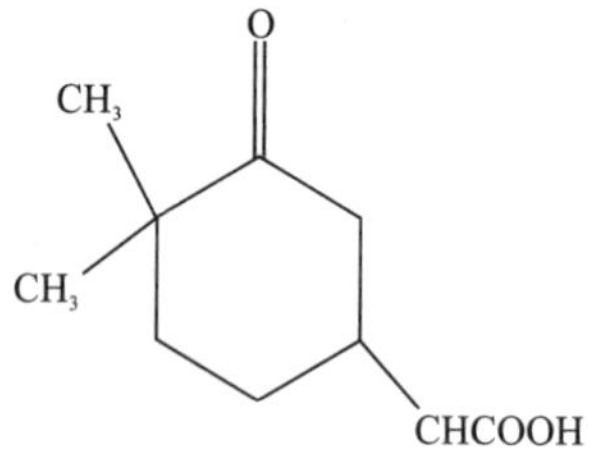

(Michael addition)

19. a. The product of the Knoevenagel condensation of pentanal, $CH_3CH_2CH_2CH_2CHO$, and diethylmalonate, $H_2C(CO_2CH_2CH_3)_2$, is as follows.

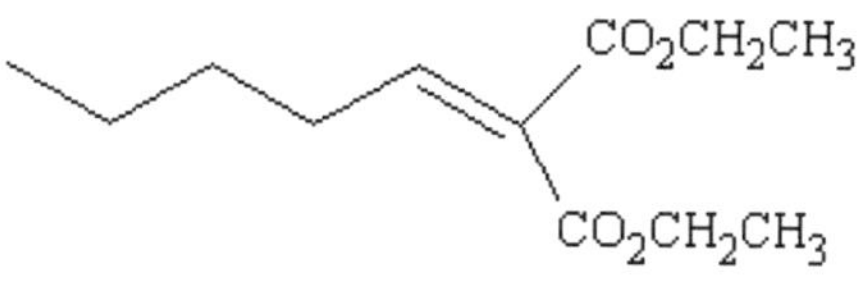

b. In this condensation reaction, a catalytic amount of a weak base such as diethylamine is heated with the reactants. **(Knoevenagel condensation)**

20. b. $C_6H_5COOCH_3$ cannot undergo a Claisen condensation because it does not have an α H atom to the carbonyl group. **(Claisen condensation)**

21. a. Methyl phenyl ketone cannot be synthesized by the acetoacetic ester synthesis because it is not an α-substituted acetone, the product of the acetoacetic ester synthesis. **(Acetoacetic ester synthesis)**

22. d. 1-Bromo-3,3-dimethylbutane, $(CH_3)_3CCH_2CH_2–Br$, undergoes the acetoacetic ester synthesis and produces 6,6-dimethyl-2-heptanone. **(Acetoacetic ester synthesis)**

23. f. The product of this reaction is octanoic acid. When *n*-hexylmalonic acid is heated, it undergoes a decarboxylation reaction, removing one of the two carboxyl groups. Thus, the six C atoms from the *n*-hexyl group and the two remaining from malonic acid yields octanoic acid. (**Malonic ester synthesis**)

24. a. Methyl and *n*-octyl groups must be added to malonic ester to produce 2-methyldecanoic acid. The remaining two C atoms were originally part of the malonic ester. (**Malonic ester synthesis**)

25. a. The Claisen condensation is used to produce β-keto esters. In this reaction, the ethyl phenylacetate produces ethyl 3-oxo-2,4-diphenylbutanoate, which can then add allyl bromide. The product of that reaction is hydrolyzed and then decarboxylated to the desired product. (**Claisen condensation**)

26. f. A Knoevenagel synthesis is the only reasonable way to do this synthesis. 3-Pentanone reacts with diethyl malonate in piperidine to yield this product. (**Knoevenagel condensation**)

27. d. Three different enols of ethyl acetoacetate can form. They are represented as follows. (**Acetoacetic ester**)

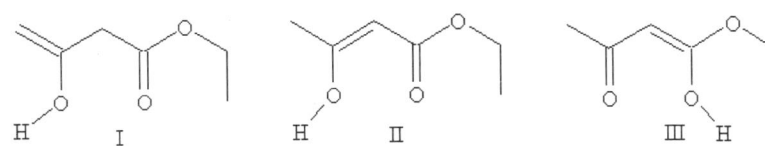

28. c. Diethyl hexanedioate undergoes the Dieckmann condensation because this reaction most readily produces five- and six-membered rings. The other listed diesters produce rings that are either too small or too large. (**Dieckmann condensation**)

29. a. 1-Phenyl-1,5-hexanedione is the product of this reaction. Ethyl acetoacetate undergoes a Michael reaction with phenyl vinyl ketone, $C_6H_5–CO–CH=CH_2$ in base and produces the following compound.

After saponification, base hydrolysis, and decarboxylation 1-phenyl-1,5-hexanedione results.

1-phenyl-1,5-hexanedione

(**Michael addition**)

30. b. Two molecules of $C_6H_5–CH_2CH_2COOCH_3$ undergo a Claisen condensation to the desired product. (**Claisen condensation**)

Grade Yourself

Circle the number of questions you missed, then fill in the total incorrect for each topic. If you answered more than three questions incorrectly, you need to focus on that topic. (If a topic has less than three questions and you had at least one wrong, we suggest you study that topic also. Read your textbook, a review book, or ask your teacher for help.)

Subject: Reactions of β-Dicarbonyl Compounds

Topic	Question Numbers	Number Incorrect
β-Keto acids	1	
Claisen condensation	2, 20, 25, 30	
Dieckmann condensation	3, 28	
Mixed Claisen condensation	4, 5	
Decarboxylation of β-keto esters	6	
Acetoacetic estersynthesis	7, 8, 21, 22	
Malonic ester synthesis	9, 10, 11, 12, 23, 24	
Michael addition	13, 14, 18, 29	
Robinson annulation	15, 16	
Reformatsky reaction	17	
Knoevenagel condensation	19, 26	
Acetoacetic ester	27	

Amines

20

Brief Yourself

Amines are organic derivatives of ammonia, NH_3. If one, two, or three R or Ar groups replace the H atoms of ammonia, primary amines, RNH_2; secondary amines, R_2NH; and tertiary amines, R_3N, result, respectively. Because of the lone pair of electrons on the N atom, a fourth R group can bond, forming a quaternary ammonium ion, R_4N^+.

The common names of amines are derived from the alkyl groups bonded to the nitrogen atom. For example, CH_3NH_2, $(CH_3)_2NH$, and $(CH_3)_3N$ are methylamine, dimethylamine, and trimethylamine, respectively. In more complicated structures, the $-NH_2$ group is called an *amino* group. Consider the compound *cis-*,4-amino-1,2-dichlorocyclohexane.

In the IUPAC system of nomenclature, the *e* from the parent alkane is removed and *amine* is added. A number is placed in front of the name to show the position of the amino group. For secondary amines, the name of the second alkyl group is placed after the prefix *N-*. For tertiary amines, the names of the second and third alkyl groups are placed after the prefix *N,N-*. Consider the following examples of IUPAC names of amines.

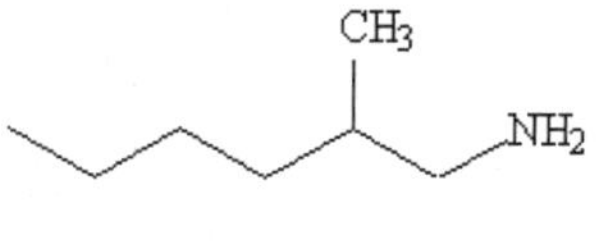

2-methylhexanamine 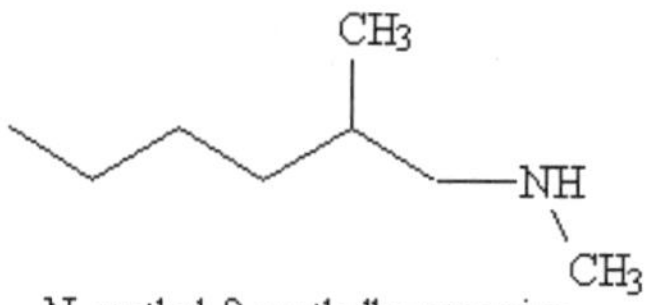   N-methyl-2-methylhexanamine 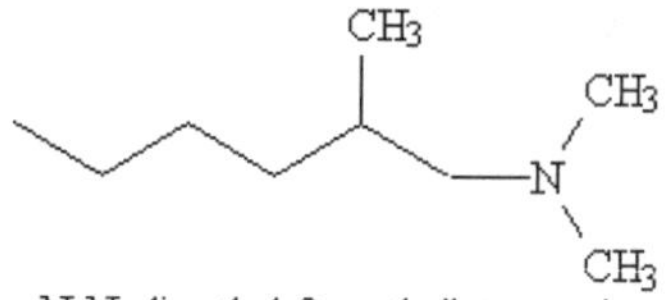   N,N-dimethyl-2-methylhexanamine

When an amino group is bonded to a benzene ring, the names are derived from the parent aromatic amine, aniline. Consider the structure of *o*-ethylaniline.

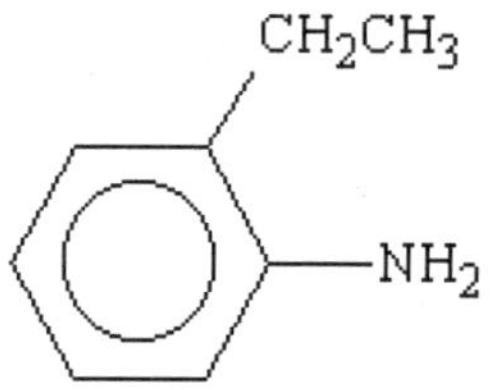

o-ethylaniline

Heterocyclic amines are compounds that have one or more N atoms in a ring. Some important heterocyclic amines are pyrrole, pyrrolidine, pyridine, and pyrimidine.

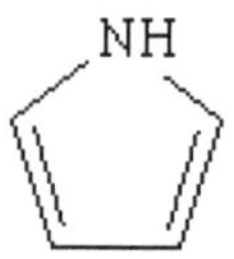

pyrrole 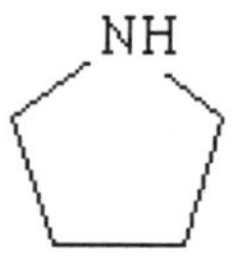   pyrrolidine 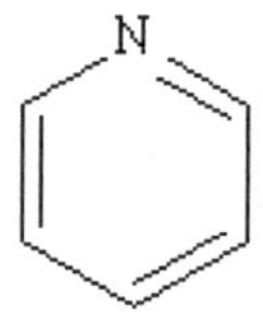   pyridine 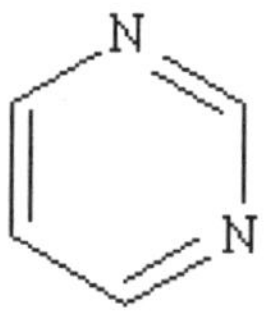   pyrimidine

The molecular geometry around the N atom in amines is trigonal pyramidal because of the sp^3 hybridization of the N atom. Due to its low energy barrier, the N atom rapidly inverts to the opposite configuration. The transition state for this inversion has a trigonal planar geometry, which reflects the sp^2 hybridization of the N atom. Because of this rapid inversion, acyclic amines with three different groups are chiral but cannot be resolved.

Amines are polar substances with moderate dipole moments because the more electronegative N atom can withdraw electron density from C atoms. Primary and secondary amines can form H bonds. Tertiary amines cannot donate hydrogen bonds because they lack a N–H bond, but can be hydrogen bond acceptors. The N–H bond is less polar the O–H bond; hence, amines form weaker H bonds than alcohols. This means that the boiling points of amines are lower than alcohols with similar structures and molecular masses. Nonetheless the boiling points of primary and secondary amines are higher than most similar nonpolar substances. Most low-molecular-mass primary and secondary amines are water-soluble.

Amines are one of the most basic groups of organic compounds. They have a lone pair on the N atom that can accept a proton, forming an alkyl ammonium salt.

$$RNH_2 + H^+ \rightleftarrows RNH_3^+$$

Amines are usually stronger bases than ammonia because alkyl groups donate electron density to N atoms. This stabilizes the resulting conjugate acid, RNH_3^+. Aromatic amines are generally weaker bases than aliphatic amines because of the resonance stabilization of the lone pair on the N atom by the delocalized π system on the benzene ring. An electron-donating substituent on an aniline ring, such as the methoxy group, makes the compound more basic, and an electron-withdrawing substituent, such as the nitro group, makes the compound less basic.

One good method for preparing amines is through reductive amination. To prepare some primary amines, an aldehyde or ketone is first reacted with hydroxyl amine, H_2NOH, producing an oxime, which is then reduced with $LiAlH_4$.

$$RCHO + H_2NOH \rightarrow RCH=NOH + LiAlH_4 \rightarrow RCH_2NH_2$$

To produce a secondary amine by this method, a primary amine and $NaBH_3CN$ is used in place of hydroxyl amine and $LiAlH_4$.

$$RCHO + R'NH_2 \rightarrow RCH=NR' + NaBH_3CN \rightarrow RCH_2NHR'$$

To produce a tertiary amine by reductive amination, a secondary amine is used in place of a primary amine.

$$RCHO + R'_2NH \rightarrow RCH=NR_2' + NaBH_3CN \rightarrow RCH_2NHR_2'$$

Another successful method for preparing amines is through acylation-reduction. In this method an amine is first reacted with an acid chloride or similar compound, producing an amide, $RCONHR'$. This is followed by reduction to the amine with $LiAlH_4$.

$$RNH_2 + R'COCl \rightarrow R'CONHR + LiAlH_4 \rightarrow R'CH_2NHR$$

In this reaction, ammonia is converted to a primary amine. Primary and secondary amines are converted to secondary and tertiary amines, respectively.

Nitro compounds can be converted to amines by reduction. Typically, an alkyl nitro compound, RNO_2, or aryl nitro compound, $ArNO_2$, is reduced with H_2 on a metal catalyst such as Ni, Pd, or Pt. Sometimes active metals such as Sn are used as catalysts.

$$RNO_2 + H_2/Ni \rightarrow RNH_2$$

Primary amines can be prepared through S_N2 reactions of alkyl halides or alkyl tosylates and excess ammonia.

$$RX + NH_3(excess) \rightarrow RNH_2 + NH_4^+X^-$$

Primary amines are also produced by the reduction of nitriles by either H_2 and a catalyst or $LiAlH_4$.

$$RCN + H_2(cat.) \rightarrow RCH_2NH_2$$

A third method that exclusively produces primary amines is the Gabriel synthesis in which phthalimide is first reacted with a strong base to produce the resonance-stabilized phthalimide anion, which is a strong nucleophile.

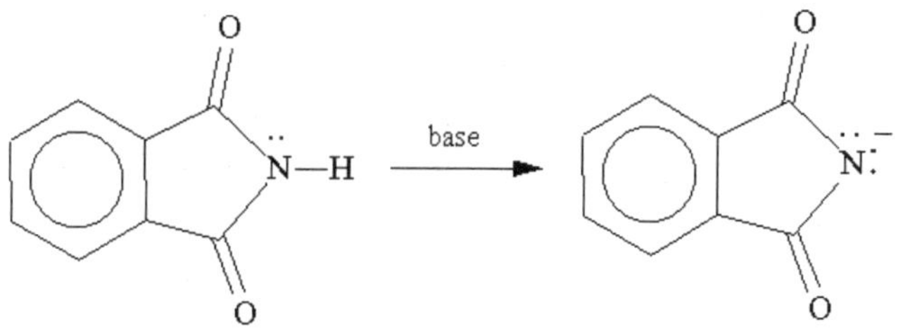

phthalimide   phthalimide anion

The phthalimide anion does an S_N2 displacement of alkyl halides or alkyl tosylates to produce *N*-alkyl phthalimides. This is followed by treatment with hydrazine, H_2NNH_2, which produces a primary amine and releases phthalimide hydrazide.

Alkyl azides, R–N_3, can be reduced to amines using either catalytic hydrogenation or $LiAlH_4$ in diethylether. In this synthesis, an alkyl halide is treated with sodium azide, NaN_3, producing the alkyl azide, which is then reduced. The following shows how cyclopentylamine could be synthesized from cyclopentylbromide.

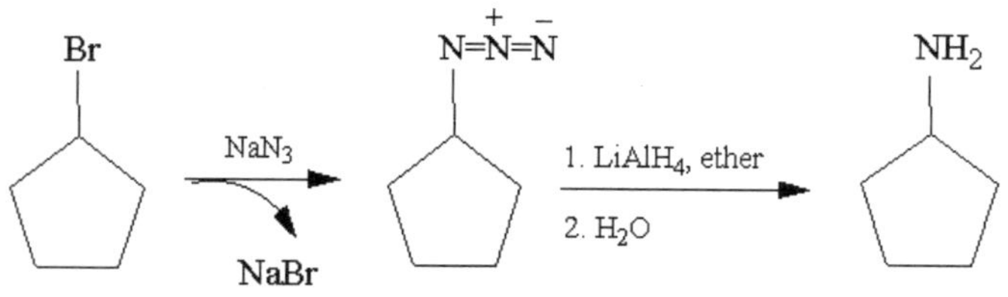

Amines are reactive and undergo many reactions. Because they are bases, amines react with acids and form alkyl ammonium salts.

$$RNH_2 + HX \rightarrow RNH_3^+ \, X^-$$

They react with aldehydes and ketones and form imines, RCH=NH, (discussed in Chapter 15). Amines are good nucleophiles and thus react with alkyl halides and alkyl tosylates, producing salts of secondary amines.

$$RNH_2 + R'CH_2X \rightarrow RNH_2^+-CH_2-R' \, X^-$$

Additionally, multiple alkylations often occur.

Amines react with acetyl chloride, RCOCl, and yield amides.

$$R'NH_2 + RCOCl \rightarrow RCONHR' + HCl$$

They also react with sulfonyl chloride, RSO_2Cl, and yield sulfonamides, RSO_2NHR'.

$$R'NH_2 + RSO_2Cl \rightarrow RSO_2NHR' + HCl$$

Amines are converted to quaternary ammonium hydroxides, $RCH_2CH_2N^+(CH_3)_3OH^-$, by first reacting with an alkyl iodide, such as CH_3I, followed by reaction with Ag_2O. When these compounds are heated, they undergo Hofmann elimination producing the least highly substituted alkene.

$$RCH_2CH_2N(CH_3)_3^+ OH^- + heat \rightarrow RCH=CH_2 + (CH_3)_3N$$

A reaction similar to the Hofmann elimination is the Cope elimination in which a tertiary amine oxide undergoes an elimination reaction and produces the least highly substituted alkene.

Primary amines react with nitrous acid (HONO), produced from $NaNO_2$ and HCl, and yield diazonium salts, $RN\equiv N^+ Cl^-$, which are unstable and immediately break down and produce carbocations, R^+, and N_2.

$$RNH_2 + HONO \rightarrow RN\equiv N^+ \rightarrow R^+ + N_2$$

Nitrous acid reacts with secondary amines and produces *N*-nitrosamines, $R_2N–NO$. Tertiary amines react with HONO and produce *N*-nitrosoammonium salts.

Unlike alkyl diazonium salts, aromatic diazonium salts, $Ar–N\equiv N^+ X^-$, are stable in aqueous solutions. Nonetheless, these compounds are reactive and many different groups can replace the diazonium ion and substitute on the aromatic ring. For example, H_2O, KI, CuCl, CuBr, CuCN, and H_3PO_2 yield Ar–OH, Ar–I, Ar–Cl, Ar–Br, Ar–CN, and Ar–H, respectively.

Test Yourself

1. What are the three general classes of amines? Write the general formula for each.

2. a. Draw the structure of *N*-methylaniline.

   b. In what class of amines does this compound belong?

3. a. Draw the structure of *cis*-1,3-cycloheptanediamine.

   b. In what class of amines does this compound belong?

4. a. Draw the structure of *N*-ethyl-*N*-isopropyl-*n*-hexylammonium acetate.

   b. In what class of amines does this compound belong?

5. Write the name for the following compound.

6. Write the name for the following compound.

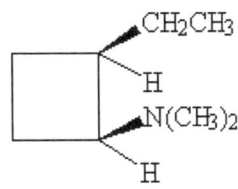

7. Explain why some quaternary ammonium salts exist as enantiomeric pairs that can be resolved, but amines with three different substituent groups cannot be resolved.

8. The pK_b values for ammonia and methylamine are 4.74 and 3.36, respectively. Explain the difference in basicity of these compounds.

9. What product(s) results when excess methyl iodide reacts with cyclopentylamine in the presence of a base in methanol? What general type of reaction is this?

10. What product results when N-methyl-(2-phenylethyl)amine is first treated with excess methyl iodide followed by Ag_2O and H_2O and the resulting product is heated? What is the name of this general type of reaction?

11. N,N-Dimethylcyclohexylamine is first treated with hydrogen peroxide in a solution of methanol and water. What product results when the resulting product of this reaction is heated? What is the name of this general type of reaction?

12. a. Draw the structure of the product that results when N,N-dimethylcyclohexylamine reacts with hydrogen peroxide (Problem 11).

    b. Show the mechanism in which this compound is converted to the final product.

13. What is the product of the reaction of piperidine with sodium nitrite and HCl in aqueous solution?

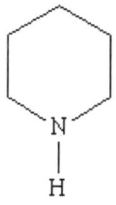

piperidine

14. Draw the structure of the product that results when piperidine (Problem 13) reacts with pentanal followed by catalytic reduction with H_2.

15. Starting with benzyl alcohol, $C_6H_5CH_2$–OH, show how the following compound can be synthesized.

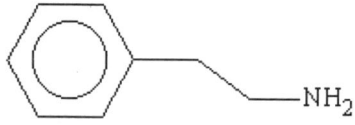

16. What amine results when isobutyl bromide is used in a Gabriel synthesis of amines?

17. *m*-Chloroaniline is first treated with sodium nitrite and hydrochloric acid. The product of this reaction is heated with KI.

    a. Draw the structure of the product of this reaction.

    b. What is the structure of the intermediate compound?

18. Consider the compounds piperidine and pyridine.

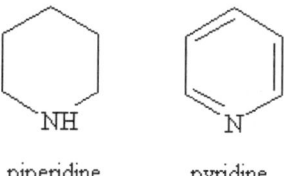

piperidine   pyridine

The pK_b values for these compounds are 8.75 and 2.88. Match the pK_b values to the correct compound and write an explanation that explains why one is significantly more basic than the other.

19. Starting with $HOOCCH_2CH_2COOH$ show how $H_2NCH_2CH_2CH_2CH_2NH_2$ is synthesized.

20. Which of the following has the highest boiling point?

    a. *n*-propylamine

    b. methylamine

    c. trimethylamine

    d. ethylamine

21. Which of the following is most basic in aqueous solution?
    a. ammonia
    b. ethylamine
    c. diethylamine
    d. triethylamine
    e. all are the same basicity

22. Which of the following is the strongest base?
    a. *p*-methyoxyaniline
    b. *p*-nitroaniline
    c. *p*-chloroaniline
    d. *p*-cyanoaniline
    e. all are the same basicity

23. What is the principal product when the following compound is heated?

    a. methylenecyclohexane
    b. 1-methylcyclohexene
    c. 3-methylcyclohexene
    d. 1-methylcyclohexane
    e. none of these

24. Which of the following results when the following compound first reacts with $NaNO_2$ and HCl followed by hypophosphorous acid, H_3PO_2?

    a. *o*-nitrotoluene
    b. *o*-aminotoluene
    c. *o*-phosphotoluene
    d. aniline
    e. no reaction
    f. none of these

25. Which of the following compounds undergoes reduction followed by hydrolysis and produces *n*-butylamine?
    a. CH_3CH_2CN
    b. $CH_3CH_2CH_2CH_2NHCH_3$
    c. $CH_3CH_2CH_2CONH_2$
    d. $CH_3CH_2CH_2CONHCH_3$
    e. none of these

26. What is the name of the following heterocyclic molecule?

    a. purine
    b. pyrimidine
    c. piperidine
    d. indole
    e. none of these

27. Which of the following is **not** a correct statement regarding the spectroscopy of amines?
    a. In the IR spectrum of amines, a strong peak is found between 3200 and 3500 cm^{-1}.
    b. In the ^1H NMR spectrum of amines, the chemical shifts of the N–H protons depend on the degree of hydrogen bonding.
    c. In the ^1H NMR spectrum of amines, the chemical shifts of N–H protons are in the range of 1 to 4 ppm (δ).
    d. In the ^{13}C NMR spectrum of amines, the chemical shifts of the C atom α to the N atom are in the range of 40 to 50 ppm (δ).
    e. all are correct

28. Aniline first reacts with acetyl chloride producing Compound *K*. *K* reacts with a nitric acid/sulfuric acid mixture and produces Compound *L*, which hydrolyzes to Compound *M*. What is the identity of *M*?

    a. acetanilide

    b. *p*-nitroacetanilide

    c. *p*-nitroaniline

    d. aniline

    e. none of these

29. Starting with acetanilide, which of the following sequences of reactions would produce sulfanilamide?

acetanilide

sulfanilamide

    a. 1. CH_3COCl, 2. NH_3/H_2O, 3. heat

    b. 1. $HOSO_2Cl$, 2. CH_3NH_2/H_2O, 3. dilute HCl/heat

    c. 1. H_2SO_4, 2. NH_3/H_2O, 3. dilute HCl/heat

    d. 1. $HOSO_2Cl$, 2. NH_3/H_2O, 3. dilute HCl/heat

    e. none of these

30. Compound *O* reacts with diethylamine and produces Compound *P*. *P* reacts with $LiAlH_4$ and produces benzyldiethylamine. What is the identity of *O*?

    a. benzoic acid

    b. benzyl alcohol

    c. phenol

    d. benzaldehyde

    e. benzophenone

    f. none of these

31. What is the product of the reaction of pentanamide with Br_2 and NaOH?

    a. $CH_3CH_2CH_2CH_2NCO$

    b. pentanamine

    c. butanamine

    d. $CH_3CH_2CH_2CH_2CONHBr$

    e. none of the above

32. Toluene first reacts with a mixture of nitric acid and sulfuric acid. The product of this reaction is treated with Sn and HCl followed by OH⁻. What is the major product of these reactions?

    a. *p*-nitrotoluene

    b. aniline

    c. *p*-methylaniline

    d. 1-amino-4-methylcyclohexane

    e. none of these

33. Compound *Q* is chiral and has the formula $C_8H_{11}N$. *Q* releases N_2 when it reacts with nitrous acid, and it dissolves in aqueous HCl. What is the identity of *Q*?

    a. 1-cyclohexylethanamine

    b. 1-phenylethanamine

    c. 2-phenylethanamine

    d. *N*-ethylaniline

    e. none of these

34. Cyclohexanone reacts with aniline and H$^+$ and produces Compound *R*. When *R* is treated with LiAlH$_4$, it produces Compound *S*. What is the identity of *S*?

    a. *N*-phenylcyclohexylamine

    b. *N*-phenylaniline

    c. *N,N*-diphenylcyclohexylamine

    d. dicyclohexylamine

    e. none of these

35. Compound *T* reacts with NaCN producing compound *U*, which is catalytically reduced to (*S*)-2-methyl-1-butanamine. What is the identity of *T*?

    a. (*S*)-2-bromobutane

    b. (*S*)-1-bromobutane

    c. (*R*)-2-bromobutane

    d. (*R*)-1-bromobutane

    e. none of these

✔ Check Yourself

1. The three general classes of amines are primary, RNH2; secondary, R2NH; and tertiary, R3N. **(Classes of amines)**

2. a. The structure of *N*-methylaniline is as follows.

N-methylaniline

   b. It is a secondary amine. **(Amine structures)**

3. a. The structure of *cis*-1,3-cycloheptanediamine is as follows.

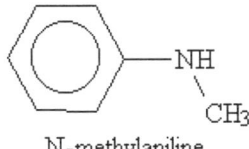

   b. It is a primary diamine because only one R group is attached to each N atom. **(Amine structures)**

4. The structure of *N*-ethyl-*N*-isopropyl-*n*-hexylammonium acetate is as follows.

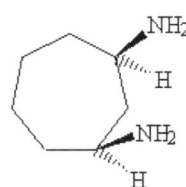

   b. It is an alkyl ammonium salt. **(Amine structures)**

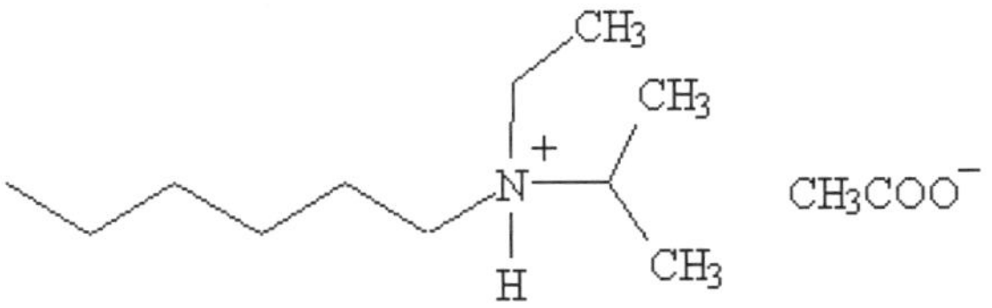

5. Correct names for this compound are *cis*-2-ethyl-*N,N*-dimethylcyclobutylamine, *cis*-2-ethyl-*N,N*-dimethylcyclobutanamine, or *cis*-2-ethyl-1-dimethylaminocyclobutane. **(Amine nomenclature)**

6. The name of this compound is *p*-nitroaniline. (**Amine nomenclature**)

7. The geometry of quaternary ammonium salts is tetrahedral; thus, those with four different groups are nonsuperimposable on their mirror images. This means they can be separated into their *R* and *S* forms. Tertiary amines with three different groups are chiral but the N atom undergoes rapid pyramidal inversion, making resolution impossible. (**Amine structures**)

8. The higher value for the pK_b of ammonia means it is a weaker base than methylamine. Bonded to the N atom in ammonia are three H atoms which do not help stabilize the resulting ammonium ion. Bonded to the N atom in methylamine is an electron-releasing methyl group that stabilizes the methylammonium ion. Greater stabilization shifts the equilibrium to the right, resulting in a stronger base. (**Amine properties**)

9. Amines are nucleophiles that undergo nucleophilic substitution reactions with alkyl halides. Therefore, cyclopentylamine undergoes an S_N2 reaction with methyl iodide to produce *N*-methylcyclopentylamine, which can substitute for another I atom producing *N,N*-dimethylcyclopentylamine. Finally, this compound reacts with methyl iodide producing trimethylcyclopentylammonium iodide. (**Amine reactions**)

10. This is a Hofmann elimination reaction. The methyl iodide forms a quaternary ammonium salt that reacts with aqueous silver oxide to produce a quaternary ammonium hydroxide, $C_6H_5–CH_2CH_2N(CH_3)_3^+$. When this compound is heated, it eliminates trimethylamine and forms styrene, $C_6H_5–CH=CH_2$. (**Hofmann elimination reaction**)

11. This is an example of the Cope elimination reaction. Initial treatment with hydrogen peroxide produces *N,N*-dimethylcyclohexylamine *N*-oxide. Upon heating, $(CH_3)_2NOH$ is eliminated and cyclohexene results. (**Cope elimination reaction**)

12. a. The product of this reaction is as follows.

b. The mechanism for the Cope elimination is concerted.

(**Cope elimination reaction**)

13. The product of the reaction is *N*-nitrosopiperidine.

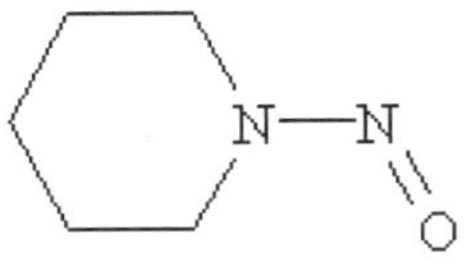

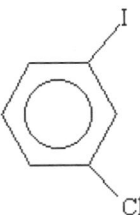

N-nitrosopiperidine

(Amine reactions)

14. Aldehydes and ketones undergo reductive amination reactions with secondary amines, such as piperidine, producing tertiary amines. The product of this reaction is as follows.

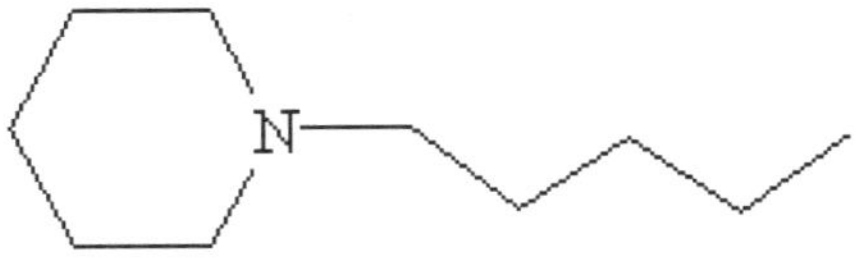

(Reductive amination)

15. First, convert benzyl alcohol to benzyl bromide, $C_6H_5-CH_2Br$, using HBr. Then treat the benzyl bromide with CN^-, producing $C_6H_5-CH_2CN$. Finally, reduce $C_6H_5-CH_2CN$ with $LiAlH_4$. The desired product results when the reduction product is hydrolyzed. **(Amine syntheses)**

16. In the Gabriel synthesis, the potassium salt of phthalimide undergoes an S_N2 reaction with a primary alkyl halide followed by hydrazinolysis. The product of the Gabriel synthesis is a primary amine. Thus, the product of this reaction is isobutyl amine, $(CH_3)_2CHCH_2NH_2$. **(Amine syntheses)**

17. a. Aniline derivatives react with $NaNO_2$ and HCl and produce aryl diazonium ions, which react with KI and produce aryl iodides. Hence, the product of this reaction is 1-chloro-3-iodobenzene.

b. The diazonium ion that results is as follows.

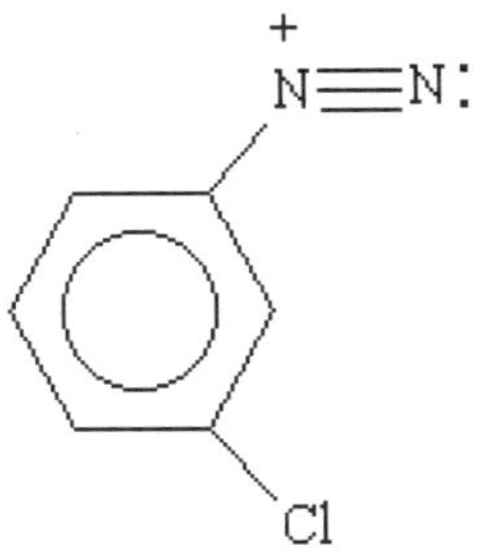

(Diazonium syntheses)

18. The pK_b of pyridine and piperidine are 8.75 and 2.88, respectively. Piperidine is a significantly stronger base because the lone pair electrons on its N atom reside in an sp^3 hybridized orbital and can be more readily donated than the lone pair electron on the N atom in pyridine which is sp^2 hybridized. Electrons in sp^2 hybrid orbitals have greater s character and are held tighter than those in sp^3 hybrid orbitals with lower s character. (**Amine properties**)

19. First, react $HOOCCH_2CH_2COOH$ with $SOCl_2$ to produce $ClOOCCH_2CH_2COOCl$. Then react this diacyl chloride with excess NH_3 to produce $H_2NOCCH_2CH_2CONH_2$, which can be reduced with $LiAlH_4$ followed by hydrolysis to $H_2NCH_2CH_2CH_2CH_2NH_2$. (**Amine syntheses**)

20. a. *n*-Propylamine has the highest boiling point. Primary and secondary amines can form intermolecular hydrogen bonds. Tertiary amines do not have a H atom bonded to a N atom to form a hydrogen bond. *n*-Propylamine is a larger molecule than either methylamine or ethylamine; thus, it will have stronger London forces. (**Amine properties**)

21. c. Diethylamine is the most basic in aqueous solution. Disubstituted amines are more basic than primary amines because they have two electron-releasing groups which stabilize the resulting conjugate acid (an ammonium ion). Trisubstituted amines are less basic than disubstituted amines because their resulting conjugate acids are not hydrated to the same degree as the conjugate acids of disubstituted amines. (**Amine properties**)

22. a. *p*-Methoxyaniline is the strongest base because the methoxy group, $CH_3O–$, is an electron-releasing group that helps stabilize the resulting conjugate acid. All of the others have electron-withdrawing groups bonded to the aromatic ring. (**Amine properties**)

23. a. Methylenecyclohexane is the principal product because in the Hofmann elimination reaction the least substituted product forms in greatest amount. (**Hofmann elimination reaction**)

24. f. The product of the reaction is toluene. The $NaNO_2$ and HCl produce the diazonium salt of *o*-toluidine. When diazonium salts are treated with H_3PO_2, the diazonium group is replaced with a H atom. This is called a reductive deamination. (**Diazonium syntheses**)

25. c. $CH_3CH_2CH_2CONH_2$ is an amide that can be reduced with $LiAlH_4$ to *n*-butylamine. (**Amine syntheses**)

26. a. The name of this heterocyclic amine is purine. (**Heterocyclic amines**)

27. a. In the IR spectrum of *primary and secondary* amines, a strong peak is found between 3200 and 3500 cm^{-1}. Tertiary amines do not have the N–H stretch. (**Amine spectroscopy**)

28. c. *M* is *p*-nitroaniline. Aniline first reacts with acetyl chloride producing acetanilide *(K)* (see Question 29). Acetanilide reacts with a nitric acid/sulfuric acid mixture and produces *p*-nitroacetanilide *(L)*, which hydrolyzes to *p*-nitroaniline. (**Amine reactions**)

29. d. When acetanilide is treated with $HOSO_2Cl$, it adds a SO_2Cl group to the para position of the ring. This is the chlorosulfonation reaction. Treatment with NH_3 and H_2O produces the sulfonamide, which upon heating with dilute HCl removes the acetyl group and produces sulfanilamide. (**Amine syntheses**)

30. f. To produce *N,N*-disubstituted amides in good yield, acid chlorides are principally used. Thus, *O* is benzoyl chloride, C_6H_5COCl. (**Amine syntheses**)

31. c. Butanamine is the product of the Hofmann rearrangement of amides. Primary amides react with a halogen and base producing an amine with one less C atom than the original primary amide. **(Hofmann rearrangement)**

32. c. Toluene is nitrated when it reacts with HNO_3/H_2SO_4, forming *p*-nitrotoluene as the major product and *o*-nitrotoluene, which must be separated. Treatment with Sn and HCl followed by OH⁻ produces *p*-methylaniline. **(Amine syntheses)**

33. b. The formula indicates that *Q* has a benzene ring because of the four degrees of unsaturation. Because *Q* is chiral it has a C atom with four different groups bonded. By releasing N_2 when treated with HONO, it is a primary amine. By dissolving in aqueous HCl, it also shows that it is an amine. The only structure consistent with this information is 1-phenylethanamine, C_6H_5–$CH(NH_2)CH_3$. **(Amine identification)**

34. a. In the presence of acid, cyclohexanone reacts with aniline and produces the following iminium salt.

This salt is reduced to *N*-phenylcyclohexylamine by $LiAlH_4$. **(Amine syntheses)**

35. c. *T*, (*R*)-2-bromobutane, undergoes an S_N2 reaction (inversion of configuration) with CN⁻ to produce (*S*)-2-cyanobutane, which reduces to (*S*)-2-methyl-1-butanamine. **(Amine syntheses)**

Grade Yourself

Circle the number of questions you missed, then fill in the total incorrect for each topic. If you answered more than three questions incorrectly, you need to focus on that topic. (If a topic has less than three questions and you had at least one wrong, we suggest you study that topic also. Read your textbook, a review book, or ask your teacher for help.)

Subject: Amines

Topic	Question Numbers	Number Incorrect
Classes of amines	1	
Amine structures	2, 3, 4, 7	
Amine nomenclature	5, 6	
Amine properties	8, 18, 20, 21, 22	
Amine reactions	9, 13, 28	
Hofmann elimination reaction	10, 23	
Cope elimination reaction	11, 12	
Reductive amination	14	
Amine syntheses	15, 16, 19, 25, 29, 30, 32, 24, 35	
Diazonium syntheses	17, 24	
Heterocyclic amines	26	
Amine spectroscopy	27	
Hofmann rearrangement	31	
Amine identification	33	

Also Available